Managing Risk and Reliability of Process Plants

Managing Risk and Reliability of Process Plants

Editor

Aanad Mishra

scitus
academics

Managing Risk and Reliability of Process Plants
Edited by **Aanad Mishra**

Printed in 2017

ISBN: 978-1-68117-413-6

Library of Congress Control Number: 2015941620

© 2016 by

SCITUS Academics LLC,
616, Corporate Way, Suite 2, 4766,
Valley Cottage, NY 10989

www.scitusacademics.com

Notice

Contents

Preface

There is much specialist material written about different elements of managing risks of hazardous industries, such as hazard identification, risk analysis, and risk management. Managing Risk and Reliability of Process Plants provides a systematic and integrated coverage of all these elements in sufficient detail for the reader to be able to pursue more detailed study of particular elements or topics from a good appreciation of the whole field. This book helps executives, managers and technical professionals achieve not only their current PSM goals, but also to make the transition to a broader operational integrity strategy. The book focuses on the energy and process industries- from refineries, to pipelines, chemical plants, transportation, energy and offshore facilities. This will be useful to chemical engineers who have to assess the safety and reliability of process plants, and reduce and manage their risks.

Editor

Quantitative Risk Assessment Integrated with Process Simulator for a New Technology of Methanol Production Plant using Recycled CO$_2$

Julia Di Domenico, Carlos André Vaz Jr, and Maurício Bezerra de Souza Jr

School of Chemistry, Department of Chemical Engineering, Federal University of Rio de Janeiro, Av. Athos da Silveira Ramos, 149, 21949-909 Rio de Janeiro, Brazil

ABSTRACT

The use of process simulators can contribute with quantitative risk assessment (QRA) by minimizing expert time and large volume of data, being mandatory in the case of a future plant. This work illustrates the advantages of this association by integrating UNISIM DESIGN simulation and QRA to investigate the acceptability of a new technology of a Methanol Production Plant in a region. The simulated process was based on the hydrogenation of chemically sequestered carbon dioxide, demanding stringent operational conditions (high pressures and temperatures) and involving the production of hazardous materials. The estimation of the consequences was performed using the PHAST software, version 6.51. QRA results were expressed in terms of individual and social risks. Compared to existing tolerance levels, the risks were considered tolerable in nominal conditions of operation of the plant. The use of the simulator in association with the QRA also allowed testing the risk in new operating conditions in order to delimit safe regions for the plant.

INTRODUCTION

The quantitative risk assessment (QRA) permits the quantification of the existing risks of an installation in order to provide data for their acceptance or not, as well as helping to assist decisions and prioritization of choices in order to reduce unacceptable ones [1]. It is an important tool to determine the risk due to use, transport, manipulation or storage of hazardous substances during the production. It shows the risk caused by some activity and gives relevant information about the acceptability of the activity to the competent authorities [2].

This methodology has been the focus of great attention during the last four decades. Literature presents a very large spectrum of technical guidelines ([1], [2] and [3]) and applications (e.g., ammonia plant[4] and [5]; aluminum extrusion industry [6]; electricity production process [7]; offshore blowout [8] etc.). Recent works covers improvement in terms of data, methods and software ([8], [9] and [10]). For instance, Bayesian Networks methods have been used to describe scenario cause–effect chains ([8] and [10]). In terms of future refinements of QRA, one of the directions pointed out by the literature is its integrated use with process simulation to evaluate the resilience of the process to support disturbances without becoming unsafe [10]. Additionally, a new tendency in the literature is to use QRA to perform a priori evaluation of the safety of the plant [11], instead of applying it to already existing facilities which is the case of vast majority of the previous studies.

Hence, the present work focus on new methodological aspects that are allowed by the integrated use of QRA and process design simulator: the assessment of risk in preliminary design stage of a new process and its ability to support disturbances without becoming unsafe.

The risks of a future still non existing plant are assessed here in order to evaluate its acceptability in a region. The application considered is the production of methanol via hydrogenation of chemically sequestered carbon dioxide, an environmentally sustainable process.

The environmental importance of converting a greenhouse gas into a useful fuel does not preclude this process from bringing potential risks. In spite of this, typical current studies on this process do not cover safety aspects, being the focus of attention limited to economical [12] or design aspects [13]. So, the development of the QRA is fundamental for this process as it is based on a technology that is still under improvement and, in the chemical route considered here [14], demands stringent operational conditions (high pressures and temperatures; catalytic process) and involves hazardous materials. Scarce information is available about industrial plants (e.g. Mitsui Chemicals, in Japan) that really implement this process, making its simulation mandatory for any risk analysis. Additionally, the evaluation is performed considering the simulation of different process conditions in order to delimit safe regions for the operation of the future plant.

According to [3], the elaboration of a QRA can be divided basically in five steps: (i) identification and characterization of the installation and the system to be analyzed; (ii) hazardous identification and definition of the accidental scenarios; (iii) estimation of the accidental consequences with vulnerability analysis and physical effects; (iv) estimation of the accidental frequency occurrences; and (v) estimation and evaluation of the risks. After the last step, it is necessary to analyze if the estimated risks are tolerable or not. The risk acceptability criteria assumed were based on the ones established by CETESB, which were – according to their public documentation [3] – developed after a wide survey of international criteria (United Kingdom, Netherlands, Hong Kong, Australia, United States and Sweden).

The following sections are organized as follows. Section 2 presents a description of the methanol plant. Section 3 describes the characteristics of the products involved. Section 4 presents a historical analysis. Section 5 describes the methodology used and the results for the identification and classification of hazards that could cause fatalities and/or damage to third party facilities. Section 6 presents the calculation of the physical consequences. For the simulation of these consequences and vulnerability, the software PHAST was used. Section 7 provides the calculation of the frequency of occurrence of each initiating event. Section 8 describes the calculation of Social and Individual Risks. Section 9 presents the conclusions of this study.

DESCRIPTION OF THE INSTALLATION

For this study a methanol production plant was considered. The production was based on the study of Souza et al. [15] that considered the chemical route proposed in reference [14]. Here the simulation was performed in the UNISIM DESIGN (Honeywell) ambient.

Location

The plant installation was considered to be located in Aracruz city (Southwest, Brazil). In 2010, the population was estimated at 81,746 inhabitants [16].

Fig. 1 shows the layout elaborated to the plant with the systems and internal equipment.

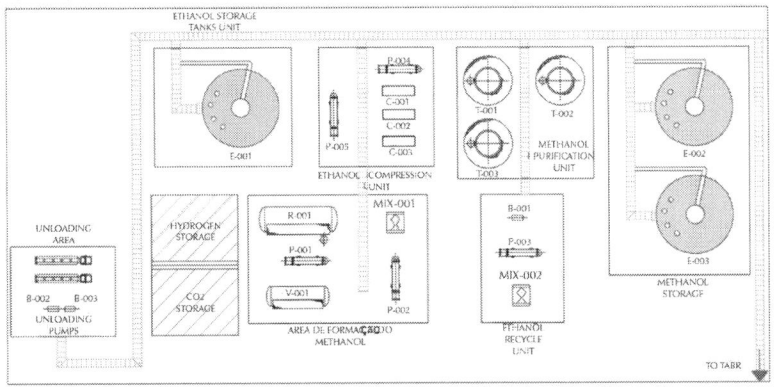

Figure 1: General layout of methanol plant.

Process Description

The methanol production process from CO_2 was based in the study of Fan at al (1998) [14]. The authors presented a production route divided in three phases, as described below. The three phases of the reaction occur at a "conversion reactor", with a pressure and temperature of 30 bar and 200 °C, respectively.

(i)   Hydrogenation of carbon dioxide (CO_2) to formic acid (HCOOH)

$$CO_2+H_2\rightarrow HCOOH \tag{1}$$

(ii) Reaction of formic acid (HCOOH), creating ethanol (C_2H_5OH) to generating ethyl formate ($HCOOC_2H_5$);

$$HCOOH+C_2H_5OH\rightarrow HCOOC_2H_5+H_2O \tag{2}$$

(iii) Production of Methanol and Ethanol through additional hydrogenolysis of ethyl formate.

$$HCOOC_2H_5+2H_2\rightarrow C_2H_5OH+CH_3OH \tag{3}$$

The process was separated in three steps, according to Fig. 2.

Step 1: methanol formation. Step 2: methanol purification. Step 3: ethanol recycle.

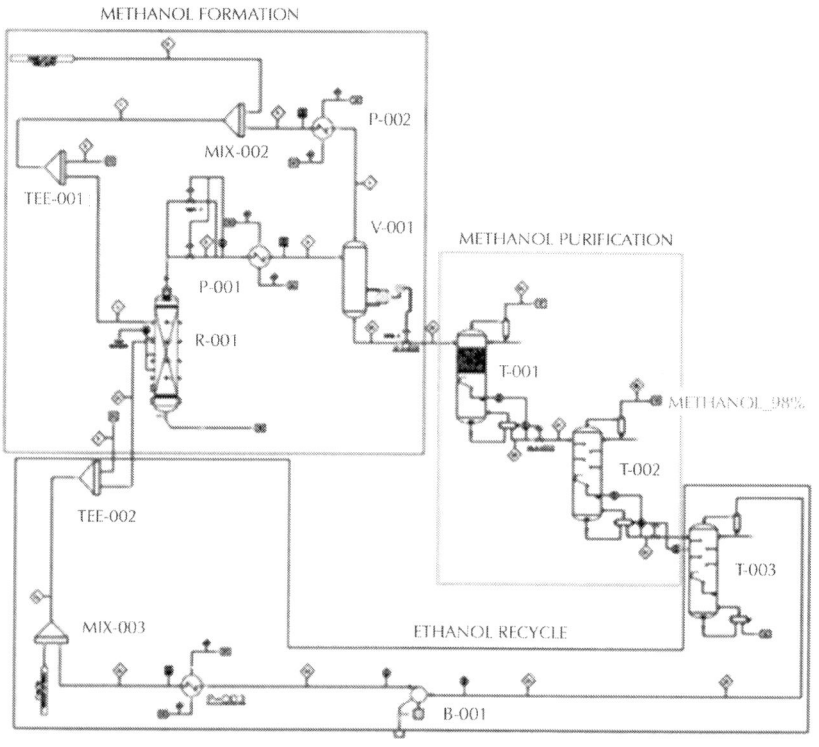

Figure 2: Methanol production steps.

Population Distribution

In this study it was decided to locate the plant in an adjacent area of Waterway Terminal of Barra do Riacho (TABR), according to Fig. 3. This region is home to three more companies and a nearby village.

Figure 3: Methanol plant location.

Meteorological Characteristics

The region meteorological data, such as temperature, air relative humidity, and atmospheric stability, are of extreme importance for atmospheric modeling dispersion and, thus, for QRA. The data were obtained from Meteorological Institute (INMET) in a period of 3 years (since 01/09/2009–01/09/2012) [17].

The atmospheric stability class is a key factor to quantitative analysis. The fact of using only characteristic meteorological data, the most used stability classification method is the Pasquill–Gifford, presented inTable 1[18].

Table 1: Classification of stability [3]

Winds velocity (V)	Day			Night	
	Insolation			Nebulosity	
	Strong	Moderate	Weak	Partial covert	Covert
V≤2	A	A–B	B	F	F

2 < V≤3	A–B	B	C	E	F
3 < V≤5	B	B–C	C	D	E
5 < V≤6	C	C–D	D	D	D
V > 6	C	D	D	D	D

*A (extremely unstable); B (unstable); C (slightly unstable); D (neutral stability); E (slightly stable); F (stable).

Table 2 consolidates the treatment of meteorological data, and presents the resulting stability class.

Table 2: Resulting meteorological data

Period	Average tempera- ture (°C)	Relative hu- midity (%)	Pres- sure (hPa)	Wind ve- locity (m/s)	Prevailing wind direc- tion	Stability class
Day	22.7	77.3	1009.1	3.3	South	C
Night	21.7	78.5	1010.1	3.5	South	D

CHEMICAL PRODUCTS INVOLVED IN THE PROCESS

The classification about the physical, toxicology and flammability state of the products used in the plant provides subsidies to the assessment of hazards and risks in the plants.

The classification of substances as to dangerousness was made using the criteria established by [3].

Among the products to be used in the plant (methanol, ethanol, formic acid, hydrogen and ethyl formate), it is worth noting the flammability of methanol, ethanol and hydrogen. The first two stand out, since their flash point is below room temperature. However, given its liquid state and low volatility, they tend not to be significant contributors to the acute fatality risk to the public outside the enterprise.

Hydrogen is presented as extremely flammable gas at standard temperature and pressure (STP). It burns at concentrations of 4% or more of H_2 in air to the upper flammability limit (85 vol%) [19]. When a leak occurs in a considerable volume of H_2, it, mixed with oxygen (O_2) explodes on ignition. The hydrogen burns violently with air with automatic ignition when reaches 400 °C [3]. Flames of pure hydrogen/oxygen are nearly invisible to the naked eye as it burns in the range of ultraviolet color. Thus, it is always necessary to install a flame detector to detect hydrogen leaks. This is considered a significant contributor to acute fatality risk to the public outside the plant.

Formic acid does not introduce significant risk of fire, since its flash point is high, and it still has a low vapor pressure (40 mmHg at 24 °C) [3].

Regarding toxicity, only formic acid was classified as toxic and dangerous according to reference [3].

HISTORICAL ANALYSIS OF ACCIDENTS (HAA)

Aiming to support and formalize the step of hazard identification, the technique HAA was used. HAA also helps to define the magnitudes associated with their consequences, allowing mainly that the accidents previously reported in similar facilities can be investigated at the concerned facility.

To elaborate the analysis, information about accidents and their frequency in similar installations were obtained from queries to databases of accidents.

The methanol plant proposal consists basically of equipment, such as tanks, vessels, towers and pumps, pipes and valves. Thus, records of accidents involving these types of equipment were considered.

The following databases were consulted for the elaboration of the analysis:

- Major Hazard Incident Data Service (MHIDAS);
- System of Information about Chemical Emergencies (SIEQ/CETESB)

The MHIDAS is an international database of accidents developed by SRD (Systems Reliability Directorate) as a representative of the main unit of the hazard assessment of the HSE (Health and Safety Executive) in the UK, being maintained by AEA Technology PLC. The MHIDAS was created in 1980, but presents registration of accidents since the beginning of the 20th century. It is a database that includes accidents in 95 countries, particularly the US, UK and Canada, where the origin of the data sources is generally public, and updated periodically [20].

It also classifies accidents into categories such as: "types of accidents", "origin", "general causes" and "affected population".

The SIEQ is a database of accidents since 1978, of the Environmental Company of the State of São Paulo (Cetesb).

For the methanol plant proposal, searches were conducted for each relevant chemical involved in the process. Table 3 displays the results of data processing for the two databases.

Table 3: Results of historical analysis of accidents

Product	MHIDAS	SIEQ
Methanol	Explosion and fire due to process or storage of unspecified cause	Accidents during road transport
Ethanol	Explosion due process by the external event or unspecified cause	Accidents during road transport
Hydrogen	Explosion equipment due to mechanical failure	Accidents during road transport
Formic acid	Fire or leak in storage due to external event	Accidents during road transport
Ethyl formate	No records of accidents involving ethyl formate	No records of accidents involving ethyl formate

HAZARDS IDENTIFICATION

In this study, the technique of preliminary risk analysis (PRA) was used. The main objective of this technique is to identify the possible risks that may be present in an industrial plant.

The classification of each individual danger in the installation is made using a qualitative categorization. These categories are adapted from the American Military Standard (MIL-STD-882) [21], along with reference [3].

In PRA, the hazards are identified with their causes and consequences, according to their respective categories of severity, frequency and combination of them (risk matrix), and also identifying some observations and recommendations.

Application of the technique

First it is selected the major subsystems of the plant to be analyzed in the PRA, as shown in Table 4. These subsystems include the entire process, and were separated according to the type of fluid, temperature, physical state and isolatable sections of the plant.

Table 4: Subsystems to be analyzed in PRA

Number	Description	Subsystem
1	Parking area	Parking lot adjacent to the trucks arrival of ethanol pipeline
2	Receiving and storage of ethanol	From receipt road ethanol, through the export pumps to storage tanks
3	Ethanol feed	From the storage tank to the mixer recycle stream ethanol
4	Recirculation and hydrogen feed	Hydrogen recirculation loop from the hydrogen cylinder to the reactor
5	Separation of hydrogen/ethanol/CO_2	From the separator vessel hydrogen/ethanol to the outlet duct of the first distillation tower (filling)

6	Production of methanol	From the inlet duct of the second distillation tower to the outlet duct of the third distillation tower
7	Storage of methanol	From the output of the distillation tower producing methanol to the storage tank
8	Recirculation of ethanol	From the outlet duct of the third distillation tower containing ethanol to the reactor through the recirculation pump
9	Methanol production reactor	Reactor

After identifying the existing hazards, a consolidation of the accidental hypotheses considered most relevant to the study of risk is done.

Fig. 4 shows the number of accidental hypotheses for each scenario. The scenarios with the highest number of accidental hypotheses selected are number 5 (separation of hydrogen/ethanol/CO_2) and number 9 (methanol production reactor). In these subsystems, there are leaks of flammable liquids and gas.

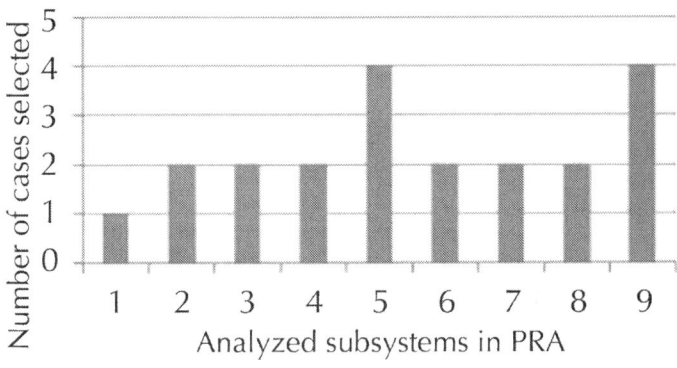

Figure 4: Relevant accidental hypothesis × scenarios.

The next phase is to characterize the various types of accidents that may occur due to uncontrolled development of events. For this study, based on the Risk Assessment Manual BEVI, 2009 [22], the types of accidents to be studied are: pool fire, explosion, flashfire, toxic cloud, fireball and jet fire. Thus the "initiating events" are found in which the occurrence requires the operation of one or more systems of security/protection so no accidents in an industrial plant occurs. The "initial event" should lead the result of accidental hypotheses generating scenarios. "Consequences" are unwanted effects of accidental scenarios. A "scenario" is an unplanned event or sequence of events that result in undesired consequence. Each scenario consists of a single primer pair initial event/consequence. If the same initial event results in different consequences, additional scenarios should be developed. In some cases, many scenarios can arise from common initiating events [23].

Table 5 shows the consolidation of cases and types of accidental initiating events (IE) to be studied in the next sections.

Table 5: Relevant initial events for QRA

Accidental hypothesis for QRA	Initial event	Description	Scenarios	Types of accidents
1.1	IE-01	Release of ethanol	1.1 A	Pool fire
			1.1 B	Flashfire
			1.1 C	Explosion
2.3	IE-03	Large release of ethanol	2.3 A	Pool fire
			2.3 B	Flashfire
			2.3 C	Explosion
3.3	IE-05	Large release of ethanol	3.3 A	Pool fire
			3.3 B	Flashfire
			3.3 C	Explosion
4.3	IE-07	Large hydrogen release	4.3 A	Fireball
			4.3 B	Explosion
			4.3 C	Flashfire
5.3	IE-09	Large release of ethanol	5.3 A	Pool fire
			5.3 B	Flashfire
			5.3 C	Explosion
5.6	IE-11	Large hydrogen release	5.6 A	Pool fire
			5.6 B	Flashfire

			5.6 C	Explosion
6.3	IE-13	Large release of methanol	6.3 A	Pool fire
			6.3 B	Flashfire
			6.3 C	Explosion
7.3	IE-15	Large release of methanol	7.3 A	Pool fire
			7.3 B	Flashfire
			7.3 C	Explosion
8.3	IE-17	Large release of ethanol	8.3 A	Pool fire
			8.3 B	Flashfire
			8.3 C	Explosion
9.3	IE-19.1	Large release of methanol/ ethanol in the reactor	9.3 A	Pool fire
			9.3 B	Flashfire
			9.3 C	Explosion
	IE-19.2	Release of formic acid in the reactor.	9.3 D	Poisoning of people
9.6	IE-21	Large release of hydrogen in the reactor	9.6 A	Fireball
			9.6 B	Explosion
			9.6 C	Flashfire

CONSEQUENCE AND VULNERA-BILITY ANALYSIS

The consequence and vulnerability analysis has the following purposes: assessment of possible damage to equipment and structures, estimation of social risk and/or individual risk for populations subject to harmful physical effects, mapping of vulnerable areas to provide grants for actions to combat emergencies, assessment of existing safeguards or prediction of additional measurements, etc.

Regulators currently base quantification of exposure to the effects of the release and dispersion of hazardous substances in death probabilities, so, only lethal effects are important. The proportion of injuries and fatalities that occur in a population exposed to thermal radiation or release of toxic substance is commonly represented by a Probit function [24]. For the simulation of the scenarios considered, the software PHAST® (Process Hazard Analysis Software Tool), version 6:51, developed by DNV Technica was used.

PHAST is a software used to assess potential danger to life and the environment, and to quantify their severity. It has the ability to simulate the progress of a potential incident from its initial leakage through the formation of a cloud or puddle until its dispersion, and automatically applying the dispersion models appropriate to the phenomenon analyzed.

In order to conduct the simulation of accident scenarios in PHAST, certain assumptions have been adopted to properly utilize the models of each result of an accident. The assumptions include:

- Probit equation used for evaluating vulnerability to thermal effects was derived by Eisenberg et al. (1975) [24] and is described as follows:

$$Pr = -14.9 + 2.56 \times \ln(Q^{4/3} \times t) \tag{4}$$

 where Pr – probit corresponding to probability of death; Q – thermal radiation (W/m^2); t – time of exposure (s).

- According to reference [3] for cases of pool fire, jet fire and fireball the time considered is 20s and 30s for the death probability of 1% and 50% respectively. Calculating the thermal radiation according to Eq. (1) the following evaluation levels shall be analyzed:
  - 3 kW/m^2 – beginning of irreversible effects;
  - 12.5 kW/m^2 – radiation with fatality probability of 1%
  - 37.5 kW/m^2 – radiation with fatality probability of 50%.
- In the case of calculating the levels of overpressure, the following equations developed by Eisenberg et al. (1975) [24] were used:

Fatality population in shelters because of structural damage:

$$Pr = -23.8 + 2.92 \times \ln(P) \tag{5}$$

Fatality of population outside shelters due to pulmonary hemorrhage

$$Pr = -77.1 + 6.91 \times \ln(P) \tag{6}$$

In Eqs. (5) and (6) P represents the pressure in Pa.

- The levels of overpressure analyzed were based on Purple Book [2], that says:
- 0.05 bar – corresponds to 1% of falling glass;
- 0.1 bar – corresponds to a fatality probability of 1% for the population in shelters;
- 0.3 bar – corresponds to a fatality probability of 50% for the population in shelters;
- For the flashfire consequence, it was considered that the concentration equivalent to the lower explosive limit (LEL) cause a fatality of 100% of the population.
- used to calculate the levels of toxicity was based on the Purple Book [2], as follows:

$$Pr = a + b \times \ln(C^n \times t) \tag{7}$$

where a, b e n – constants that describe the toxicity of the substance (specific for each substance); C – concentration (mg/m^3); t – time of exposure (min).

According to reference [3], it is recommended to use the values from the substance acrylonitrile, which are: $a = -7.52$, $b = 1$ and $n = 1.3$. Thus, the levels of physical effects for 1% and 50% fatality are 431 ppm and 2159 ppm respectively.

Vulnerabilility

From the results obtained, the maximum ranges for the delineation of vulnerable areas were determined. Table 6 presents these values for each type analyzed.

Table 6: Maximum ranges vulnerability

Initiating event	Type	Level	Fatality	Range (m)
IE-05-G night	Pool fire	12.5 kW/m^2	1%	1111

IE-05-G	Flashfire	LEL[a]	100%	188
IE-05-G night	Explosion	0.1 bar	1%	265
IE-19.2-G night	Intoxication	431	1%	1502

[a]Lower explosive limit.

Graph Mapping Results

In the vulnerability mapping, the larger distance of the radiation level of 12.5 kW/m^2 with 1% fatality was represented, as well as the level of overpressure of 0.1 bar with 1% fatality. The largest distance obtained for toxicity related to 1%, plus the distance from the LEL were also represented. Fig. 5 shows the mapping of vulnerability and Table 7 shows the legend of the mapping.

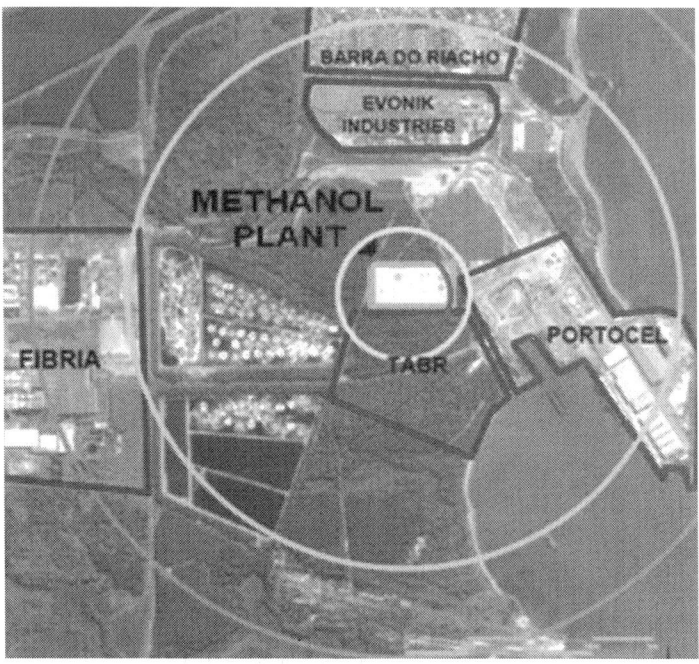

Figure 5: Vulnerability mapping.

Table 7: Legend of the vulnerability reaches

Type	Level	Fatality	Reach (m)	Color
Pool fire	12.5 kW/m^2	1%	1111	Blue
Flashfire	LEL	100%	188	Red
Explosion	0.1 bar	1%	265	Yellow
Intoxication	431	1%	1501.8	Orange

The typology "Intoxication" resulted in the greatest range observed. This event is characterized by a major leak in the reactor and reached 1502 m and 816 m to 1% and 50% of fatality respectively.

The typology "Pool Fire" and "Flashfire" resulted for some scenarios in consequences that go beyond the physical limits of the industrial area.

For the typology "Explosion" it was not verified any case that generates consequences "associated with the fatality of 50%. For the fatality of 1% almost half of the cases did not reache fatality levels. These results can be explained by the fact that the explosion cloud is directly related to the mass of product between the explosive limits on vapor cloud and the degree of confinement. It is possible to see, in the vulnerability mapping, that the consequences affect the facilities around the Methanol Plant.

FREQUENCY CALCULATION

Frequency analysis involves estimating the likelihood of each of the failure cases that were defined in the hazard identification stage. The probability of failure is defined as the probability of the event occurring in a given time period or the conditional probability of it occurring, given that a previous event has occurred [25].

One way to represent the accident frequency is using the event tree analysis (ETA). The purpose of an event tree is to present the way an accident may develop from an initiating event through several branches to one of several possible outcomes. The technique is usually used to extend the initiating event frequency estimated by one of the above means into a failure case frequency suitable for combining with the consequence models [1].

After the construction of all the event trees, the frequency of accidental scenarios linked to their consequences (pool fire, explosion, etc.) are obtained by the product of the frequency of the initiating event and the probabilities of intermediate events, such as immediate ignition, delayed ignition and wind direction. For each type of leak (liquid, gas, large, small, etc.) there is a specific event tree defined by Bevi[19].

Fig. 6 shows an example of event tree.

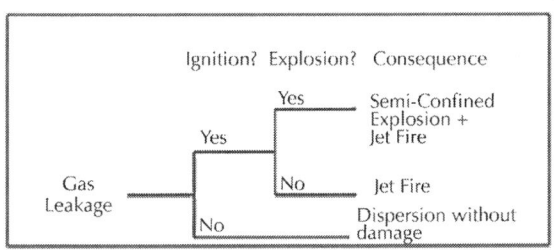

Figure 6: Example of event tree used to set consequences of failure.

For the ignition reference values of the Purple Book [2] and HSE (Health and Safety Executive), England (SRD 1978) were used. The data used for this analysis was taken from Bevi [19].

Table 8 presents the results for the frequency of occurrence of accident scenarios.

Table 8: Results of frequency of occurrences

Initiating event	Fluid leakage	Frequency (occurrence/year)	Accidental scenario	Frequency (occurrence/year)
IE-01	Ethanol	2.411E−07	Pool fire	1.567E−08
			Flashfire	1.217E−07

			Explosion	8.115E−08
IE-03	Ethanol	1.857E−05	Pool fire	1.207E−06
			Flashfire	9.378E−06
			Explosion	6.252E−06
IE-05	Ethanol	2.099E−04	Pool fire	1.364E−05
			Flashfire	1.060E−04
			Explosion	7.065E−05
IE-07	Hydrogen	1.500E−04	Fireball/jet fire	2.100E−05
			Explosion	4.680E−05
			Flashfire	7.020E−05
IE-09	Ethanol	3.370E−05	Pool fire	2.191E−06
			Flashfire	1.702E−05
			Explosion	1.134E−05
IE-11	Hydrogen	3.370E−05	Fireball/jet fire	4.718E−06
			Explosion	1.051E−05
			Flashfire	1.577E−05
IE-13	Methanol	6.300E−06	Pool fire	4.095E−07
			Flashfire	3.181E−06
			Explosion	2.121E−06
IE-15	Methanol	2.860E−05	Pool fire	1.859E−06
			Flashfire	1.444E−05
			Explosion	9.627E−06
IE-17	Ethanol	7.490E−05	Pool fire	4.869E−06
			Flashfire	3.782E−05
			Explosion	2.521E−05
IE−19.1	Methanol/ ethanol	5.000E−06	Pool fire	3.250E−07
			Flashfire	2.525E−06
			Explosion	1.683E−06
IE-19.2	Formic acid	5.000E−06	People In- toxication	5.000E−06
IE-21	Hydrogen	5.000E−06	Fireball/jet fire	7.000E−07
			Explosion	1.560E−06
			Flashfire	2.340E−06

It is noted from Table 8 that the greatest frequency found was for the initiating event IE-05, with a value of 2.099E-04 occurrences per year. This is explained because this event contains the largest number of equipment beyond a section of pipe. In this case, the equipment are two compressors, one mixer and two heat exchangers.

RISK EVALUATION

According to Crowl [25], the risk is a function that relates the frequency of occurrence of accident scenarios to consequences related to potential damage to external populations present along the designed facilities. Thus, based on the results obtained in the previous sections, it is possible to estimate the risk introduced by the methanol production plant.

In this study the risks were estimated and presented in the form of social risk and individual risk, based on comparison with the tolerance criteria recommended in reference [3]. For approval of the project both social and individual risk must be met.

Individual Risk

The individual risk is defined as the frequency of occurrence of a death due to an accident in an industrial installation related to a person who is situated at a specific location in the industry [3].

The calculation of the total individual risk in a given point can be calculated according to the Eqs.(8) and (9).

$$RI_{x,y} = \sum_{i=1}^{n} RI_{x,y,i}$$

(8)

$$RI_{x,y,i} = F_i \times p_{fi}$$

(9)

where $RI_{x,y}$ – total individual risk of fatality at the point x, y; $RI_{x,y,i}$ – fatality risk in the x, y point due to event i; n – total number of events; F_i – frequency of occurrence of a final event i; p_{fi} – probability that the event i will result in fatality at point x, y in accordance with the resulting effects from the expected consequences.

According to reference [3], the tolerance criteria are:

- Maximum tolerable risk: 1×10^{-5}/year;
- Negligible risk: $<1 \times 10^{-6}$/year.

Social Risk

The social risk is a measure of a group risk consisting of the community exposed to the effects of the accident [3].

The presentation of social risk was made by the $F{-}N$ curve, also known as "Complementary Cumulative Distribution Curve" obtained from the cumulative frequency of the accident versus the number of fatalities associated.

It is important to highlight the risk located in the region called ALARP (as low as reasonably practible). Although this region is located below the intolerable region, the risks should be reduced as much as possible.

Obtained Results

The individual risks were represented as iso-risk curves and are presented in Fig. 7. Table 9 presents the legend of individual risks.

Figure 7: Iso-risk curves.

Table 9: Individual risk legend

Level	Color
1×10^{-5}/year	Orange
1×10^{-4}/year	Green

According to Fig. 7, it is possible to note that the risk is within the limits of tolerability. The maximum level of tolerability (1×10^{-5}/ano) – the orange curve – exceeds the limits of the plant, but only reach the TABR and Portocel facilities. Levels considered tolerable only reach the interior of the facility and does not represent an external danger.

Based on the results of the frequency calculations and vulnerability occurrences of the initiating events, it is possible to draw the F–N curve to express the social risks of methanol plant, as shown in Fig. 8.

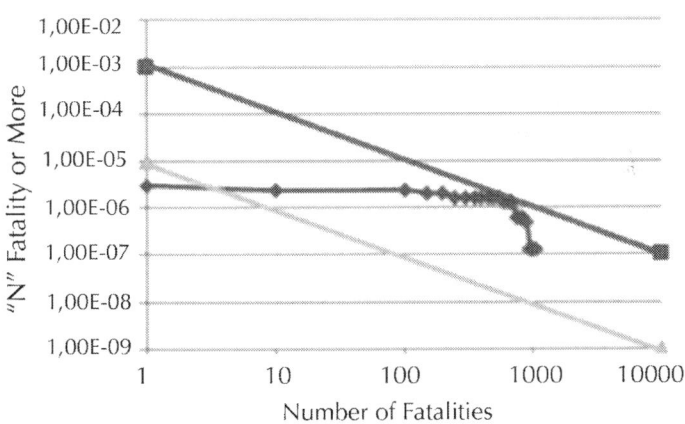

Figure 8: Social risk.

Based on the results obtained in this study, it can be seen that the social risks introduced by Methanol Plant, considering the analyzed scenarios and assumptions, lie within the negligible and ALARP regions, according to [3], which indicates that the risks introduced can be tolerated. It is important to note that, although the risk is situated within the ALARP area, it is almost at the limit of this range. It is advisable to take mitigation measures, with the incorporation of additional security features to try and reduce the risk.

The importance of the use process simulation associated with the QRA is illustrated in the following.

The results of social risk were performed under nominal conditions, chosen as the optimum working condition of the reactions [15]. In order to evaluate a more severe situation, a new simulation was performed in UNISIM DESIGN and a new state was calculated for the process considering operating conditions of 400 °C and 40 bar. In other words, a 100% higher temperature and 33% higher pressure within the reactor in comparison with the optimum conditions. Fig. 9 illustrates the result of social risk if by chance the plant work under the most extreme conditions of operation. It is possible to note that under these conditions, the social risk exceeds the ALARP region, not satisfying the tolerability criteria [3].

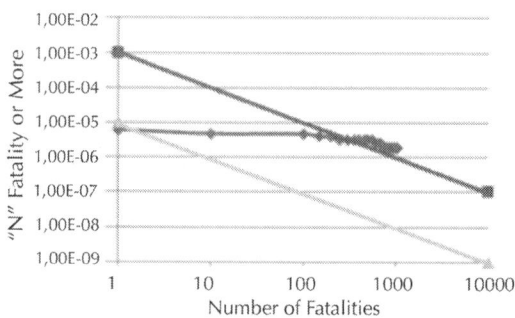

Figure 9: Social risk under extreme conditions.

CONCLUSIONS

The use of process simulators can contribute with risk analysis and assessment methods by minimizing expert time and large volume of data [26], being mandatory in the case of a non-existing future plant.

This work illustrates the advantages of this association by applying QRA to a new process: the production of methanol via hydrogenation of chemically sequestered carbon dioxide. A simulator was used both to design the process and simulate different operating conditions. The QRA allowed the risk assessment in an early design stage. So, the present work is innovative both in terms of methodology and application.

Based on the frequency and consequence results of the events it was possible to evaluate the risks of the methanol installation.

With the individual risk, as observed in the iso-risk curves, presented in Fig. 7, it was verified that the risk level of de 1×10^{-4} occurrences/year, below the maximum tolerable, is reached only in the methanol plant interior. In other words, the intolerable risks do not reach ranges outside the installation. It is also noticed that the risk level is of 1×10^{-5}/year, corresponding to the maximum tolerable level by the tolerability criteria adopted for this study, is reached, but only in a section of the TABR and Portocel.

As for the social risk imposed by the plant, a tolerable result was obtained, situated within the ALARP region. This result, however, should be further studied in order to obtain mitigation measures that reduce the risk, since it is very near the limit considered tolerable [3]. It is important to note that the results were considered satisfactory only for the nominal operation conditions of the plant [15]. Since the result of social risk bordered the tolerance limit, a second plant simulation, operating under the most extreme conditions of operation was performed. The result of social risk for this second simulation was considered unsatisfactory, as the curve F–N exceeded ALARP region.

The potential damage from potential accidents identified, as well as proximity to urban occupations in the vicinity of the installation, suggests that the risks of the enterprise must be managed and that preventive measures and/or mitigating measures presented in APR shall be evaluated and incorporated if they are appropriately viable.

REFERENCES

1. CCPS, Guidelines for Chemical Process Quantitative Risk Analysis, second ed., Centre for Chemical Process Safety, American for Chemical Engineers, New York, 2000.

2. B.J.M. Ale, P.A.M. Uijt de Haag, Guidelines for Quantitative Risk Assessment, Purple Book, third ed., Ministry of VROM, The Hague, Netherlands, 2005.

3. Cetesb, Environmental Company of São Paulo State, São Paulo, 2003, Available at: http://www.cetesb.sp.gov.br (accessed 03.04.12).

4. I.A. Papazoglou, M. Christou, Z. Nivolianitou, O. Aneziris, On the management of severe chemical accidents DECARA: a computer code for consequence analysis in chemical installations. Case study: ammonia plant, J. Hazard. Mater. 31 (1992) 135–153.

5. I.A. Papazoglou, Z. Nivolianitou, O. Aneziris, M. Christou, Probabilistic safety analysis in chemical installations, J. Loss Prevent. Proc 3 (1992) 181–191.

6. P.K. Marhavilas, D.E. Koulouriotis, A risk-estimation methodologiacal framework using quantitative assessment techniques and real accidents' data: application in an aluminium extrusion industry, J. Loss Prevent. Proc. 21 (2008) 596–603.

7. P.K. Marhavilas, D.E. Koulouriotis, Developing a new alternative risk assessment framework in the work sites by including a stochastic and deterministic process: a case study for the Greek Public Electric Power Provider, J.Safety Sci. 50 (2012) 448–462.

8. N. Khakzad, F. Khan, P. Amyotte, Quantitative risk analysis of offshore drilling operations: a Bayesian approach, Safety Sci. 57 (2013) 108–117.

9. M. Abuswera, P. Amyotte, F. Khan, A quantitative risk management framework for dust and hybrid mixture explosions, J. Loss Prevent. Proc. 26 (2013) 283–289.

10. H. Pasman, G. Reniers, Past, present and future of quantitative risk assessment (QRA) and the incentive it obtained from land-use planning (LUP), J. Loss Prevent. Proc. 28 (2014) 2–9.

11. A.M. Shariff, C.T. Leong, Inherent risk assessment – a new concept to evaluate risk in preliminary design stage, Proc. Saf. Environ. Protect. 87 (2009) 371–376.

12. F.G. Üc̦tug, S. Agrali, Y. Arikan, E. Avcioglu, Deciding between carbon trading and carbon capture and sequestration: an optimisation-based case study for methanol synthesis from syngas, J. Environ. Manage. 132 (2014) 1–8.

13. E.S. Van-Dal, C. Bouallou, Design and simulation of a methanol production plant from CO2 hydrogenation, J. Clean. Prod. 57 (2013) 38–45.

14. L. Fan, Y. Sakaiya, K. Fujimoto, Low-temperature methanol synthesis from carbon dioxide and hydrogen via formic ester, Appl. Catal. A: Gen. 180 (1999) L11–L13.

15. E.C.L. Souza, J.D.D. Pinto, V.A. Flores, Sequestro de CO2 para Produc̦ão de Metanol: Avaliac̦ão de Rotas Alternativas de Produc̦ão, Federal University of Rio de Janeiro, Rio de Janeiro, 2010.

16. IBGE, Brazilian Institute of Geology, 2012, Available at: http://www.ibge.gov.br (accessed on 28.09.12).

17. INMET, Meteorological Institute, 2012, Available at: http://www.inmet. gov.br/portal/ (accessed on 28.09.12).

18. S. Mannan, Lees' Loss Prevention in the Process Industries, Hazard Identification Assessment and Control, vol. 1, third ed., Butterworth-Heinemann, Texas, United States, 2004.

19. M.N. Carcassi, F. Fineschi, Deflagrations of H2–air and CH4–air lean mixtures in a vented multi-compartment environment, Energy 30 (2005) 1439–1451.

20. R.M. Darbra, J. Casal, Historical analysis of accidents in seaports, Department of Chemical Engineering, Polytechnic University of Catalunya, Spain, 2004.

21. Standard practice for system safety (MIL-STD-882D),

Department of Defense, United States, 2000.

22. Reference Manual Bevi Risk Assessment – RIVM, rev. 3.2, National Institute of Public Health and the Environment, Holthe, Netherlands, 2009.

23. F.M. Vasconcelos, Uma Aplicac͵ão da Técnica de Análise de Camadas de Protec͵ão (LOPA) na Avaliac͵ão do Risco do Sistema de Hidrogênio de Refrigerac͵ão do Gerador Elétrico Principal de uma Usina Nuclear, Federal University of Rio de Janeiro, Rio de Janeiro, Brazil, 2008.

24. N.A. Eisenberg, Vulnerability model. A simulation system for assessing damage resulting from marine spills, 1975, Report No. CG-D-38-79.

25. D.A. Crowl, J.F. Louvar, Chemical Process Safety: Fundamentals with Application, third ed., Prentice Hall International Series in the Physical and Chemical Engineering Sciences, Boston, United States, 2011.

26. F.I. Khan, S.A. Abasi, Techniques and methodologies for risk analysis in chemical process industries, J. Loss Prevent. Proc. 11 (1998) 261–277.

Advances in Asset Management Techniques: An Overview of Corrosion Mechanisms and Mitigation Strategies for Oil and Gas Pipelines

Chinedu I. Ossai

Production Planning Department, Overall Forge Pty Ltd, 70 R W Henry Drive, Ettamogah near Albury, Albury, NSW 2640, Australia

ABSTRACT

Effective management of assets in the oil and gas industry is vital in ensuring equipment availability, increased output, reduced maintenance cost, and minimal nonproductive time (NPT). Due to the high cost of assets used in oil and gas production, there is a need to enhance performance through good assets management techniques. This involves the minimization of NPT which accounts for about 20–30% of operation time needed from exploration to production. Corrosion contributes to about 25% of failures experienced in oil and gas production industry, while more than 50% of this failure is associated with sweet and sour corrosions in pipelines. This major risk in oil and gas production requires the understanding of the failure mechanism and procedures for assessment and control. For reduced pipeline failure and enhanced life cycle, corrosion experts should understand the mechanisms of corrosion, the risk assessment criteria, and mitigation strategies. This paper explores existing research in pipeline corrosion, in order to show the mechanisms, the risk assessment methodologies, and the framework for mitigation. The paper shows that corrosion in pipelines is combated at all stages of oil and gas production by incorporating field data information from previous fields into the new field's development process.

INTRODUCTION

The oil and gas industry is an asset intensive business with capital assets ranging from drilling rigs, offshore platforms and wells in the upstream segment, to pipeline, liquefied natural gas (LNG) terminals, and refineries in the midstream and downstream segments. These assets are complex and require enormous capital to acquire. An analysis of the five major oil and gas companies (BP, Shell, ConocoPhillips, Exxonmobil, and Total) shows that plant, property, and equipment on average accounts for 51% of the total assets with a value of over $100 billion [1]. Considering the huge investment in assets, oil and gas companies are always under immense pressure to properly manage them. To achieve this involves the use of different optimization strategies that is aimed at cost reduction and improved assets reliability [2].

Due to the growth in the demand of oil and gas around the world, companies are developing new techniques to reach new reservoirs in the offshore and onshore arena [3]. This is putting pressure on most of the facilities with the attendant cost of maintenance soaring [1]. The continuous utilization and the ageing of facilities have resulted in record failures in the oil and gas plants. Research shows that between 1980 and 2006, 50% of European, major hazards of loss containment events arising from technical plants failures were primarily due to ageing plants mechanism caused by corrosion, erosion, and fatigue [4, 5].

A study shows that corrosion cost in US rose above 1$ trillion in 2012 accounting for about 6.2% of GDP hence, the largest single expense in the economy [6]. In the oil and gas company, corrosion accounts for over 25% of assets failure [7] and is found to be prevalent in every stage of the production cycle. Oxygen which plays a dominant role in corrosion is normally present in producing formation water. During drilling operation, drilling mud can corrode the well casing, drilling equipment, pipeline, and the environment. Water and CO_2 produced or injected for secondary recovery can cause severe corrosion of completion strings, while the acids used to reduce formation damage around the well or to remove scale can attack metals [8]. The formation water and injected water used for the oil recovery are a potential source of pipeline corrosion during transportation of the oil from the wells to the loading terminals. Mechanical static equipment like valves, tanks, vessels, separators, and so forth are susceptible to a different kind of corrosion however, pipelines are more prone to corrosion due to the presence of CO_2, H_2S, H_2O, bacteria, sand, and so forth in the fluid.

Owing to the increasing cost of pipeline corrosion management in the oil and gas industries [1], operators are becoming more concerned about corrosion management planning at all phases of production. Corrosion information from existing field data is being incorporated into design information for new oil and gas field [9, 10] in a bid to develop appropriate corrosion management methodologies that will enhance the design life of the pipelines and optimize production. To reduce the risk of microbiologically influenced Corrosion (MIC) and other associated corrosions like stress corrosion cracking (SCC), hydrostatic testing of carbon steel pipes should be carried out in such a manner that enhances the future pipeline service conditions by using the right source of water, ensuring proper degree of filtration, ensuring limited exposure period to temperature and eliminating air packets [11]. Though bacteria in the biofilm are responsible for pitting of a pipeline in a MIC however, the impact of the flow velocity of the constituent fluid influences the mass transfer rate thereby affecting the biofilm formation, hence, inhibiting the activities of sulphate reducing bacteria, (SRB) present in the fluid [12]. This flow attribute has significant impact in MIC in oil and gas pipeline.

Considering the fact that the CO_2 and H_2S induced corrosion rate can reach up to 6 mm/yr and 300 mm/yr, respectively, [13] in oil and gas pipelines, sophistication in inspection and monitoring techniques is therefore necessary for quick mitigation. The increased trend in in-line inspection and online data acquisition has helped in quicker data acquisition, analysis, and decision making regarding corrosion in pipelines. The enhanced research knowledge of the behaviour of these corrodents (CO_2 and H_2S, acetic acid, etc.) at different operating conditions [14–17] has given rise to numerous mechanistic, statistical, and empirical models [18–23] which have contributed immensely in the inspection and monitoring, selection of inhibitors, and materials selection for pipelines design.

Since corrosion is a dominant factor contributing to failures and leaks in pipelines [24], to aid industry experts in managing the integrity of pipelines therefore involves a layout of the developments in the management strategies. This involves the recognition of the conditions contributing to the corrosion incident and identifying effective measures that can be taken to mitigate against them. To facilitate best practices in pipeline integrity management therefore, requires a framework that utilizes good policies and procedures in inspection, data collection, and interpretation for corrosion control.

OVERVIEW OF CORROSION

Corrosion is a naturally occurring phenomena commonly defined as the deterioration of a substance (usually metal) or its properties because of a reaction with its environment [25]. Corrosion of materials is inevitable due to the fundamental need of lowering of Gibbs energy [26]. Every material is trying to achieve a lower energy state hence the ability to corrode in order to get to a low energy oxide state. Though this is the case with all materials, the major focus of experts however, is to achieve an equilibrium position between the materials and the environment thereby controlling corrosion.

Modern corrosion science has its roots in electrochemistry and metallurgy. Whereas electrochemistry contributes to the understanding of materials via corrosion, metallurgy provides information about the behaviour of the material and their alloys hence provide a medium for combating the degradation on them. The type of corrosion mechanism and its rate of attack depend on the nature of the environment (air, soil, water, etc.) in which the corrosion takes place. Whereas some environmental condition can help to mitigate the rate of corrosion, others help to increase it hence, industrial wastes and products can either be corrosion inhibitor or catalyst. For instance, CO_2, H_2S, temperature, mass flow rate, pH, formation water, and so forth contribute in no small measure to the rate of corrosion in oil and gas pipeline [14, 16, 17, 27]. The existence of anodic cathodic sites on the surface of a piece of metal implies that the difference in electrical potential is found on the surface. This potential difference has the tendency of initiating corrosion. If an oil and gas pipeline passes through a zone of clay soil (where the oxygen concentration is low) to gravel (where the oxygen concentration is high), the part of the pipeline in contact with the clay becomes anodic and suffers damage. Though this problem is extensively addressed with the cathodic protection [26], concentration cell may also be formed where there are differences in metal ion concentration.

Although most metals are crystalline in form, they generally are not continuous single crystal but rather are collections of small grains of domains of localized order in which microcrystal forms as the liquid cools and solidifies. In the final states, the crystals have different orientation with respect to one another. The edge of the domain form grain boundaries which are an example of planar defects in metal. These defects are usually sites of chemical reactivity. The boundaries are also weaknesses, the places where stress corrosion cracking begins. The metallic surface exposed to an aqueous electrolyte usually possesses site for oxidation (anodic reaction) that produces electrons in the metal and reduction (cathodic reaction) that consumes the electrons produced by the anodic reaction [25, 26]. These sites make up a corrosion cell. The anodic reaction (Figure 1) involves the dissociation of metal to form either soluble ionic product or an insoluble compound of metal usually an oxide. For cathodic reaction (Figure 2), oxygen gas generated could be reduced or water is reduced to produce hydrogen gas. The simultaneous reaction of the anodic and cathodic reactions produces the electrochemical cell.

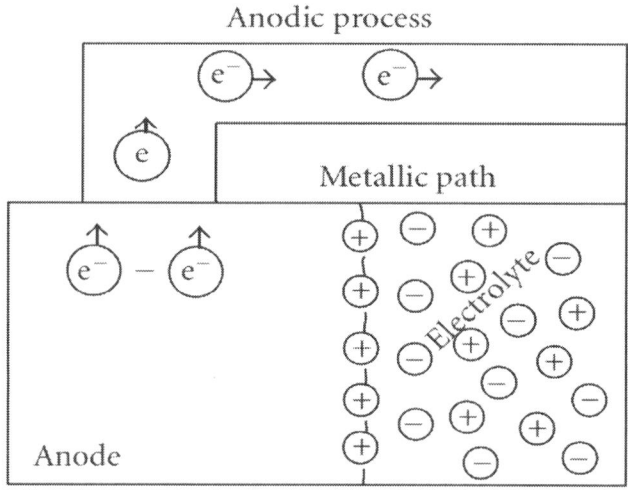

Figure 1: Anodic process.

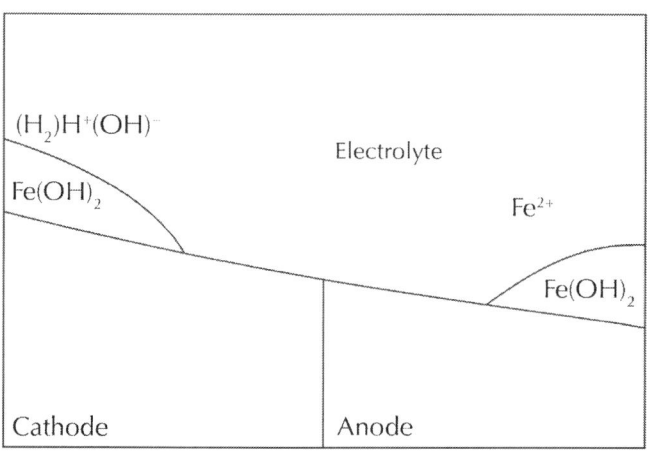

Figure 2: Cathodic process.

In completely oxygen-free water, the cathodic reaction that takes place is the reaction of hydrogen ion to form hydrogen gas as shown in (1):

$$2H^+ + 2e^- \longrightarrow H_2(g) \tag{1}$$

When significant amounts of oxygen are present in the system, the cathodic reaction that takes place is shown in (2):

$$2H^+ + \frac{1}{2}O_2 + 2e^- \longrightarrow H_2O \tag{2}$$

The hydrogen ion is present in water due to the ubiquitous dissolution of water into hydroxyl ions as shown in (3):

$$2H_2O \longrightarrow 2H^+ + 2(OH)^- \tag{3}$$

In the anode, there is a dissociation of iron to form a ferrous ion as shown in (4).

$$Fe \longrightarrow Fe^{2+} + 2e^- \tag{4}$$

The ferrous ion will react with the hydroxyl ion to form insoluble ferrous hydroxide as shown in (5):

$$Fe^{2+} + 2(OH)^- \longrightarrow Fe(OH)_2 \tag{5}$$

The anodic and cathodic reactions that take place in a neutral and alkaline condition is shown in Figure 3.

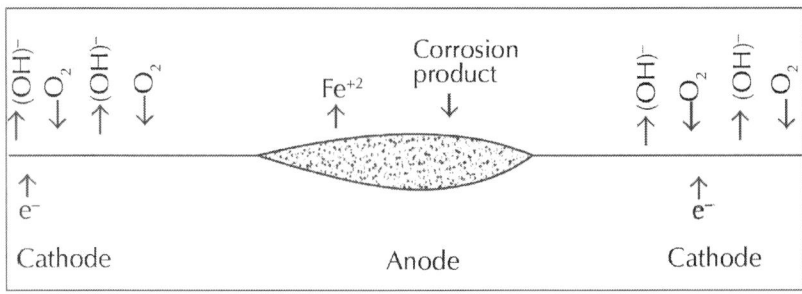

Figure 3: Neutral and Alkaline condition of a corrosion process.

The cathodic reaction is as follows:

$$\frac{1}{2}O_2 + H_2O + 2e^- \longrightarrow 2(OH)^-$$

(6)

while the anodic reaction is the same as (4).

For an anodic condition, the cathodic and anodic reactions are represented in Figure 4.

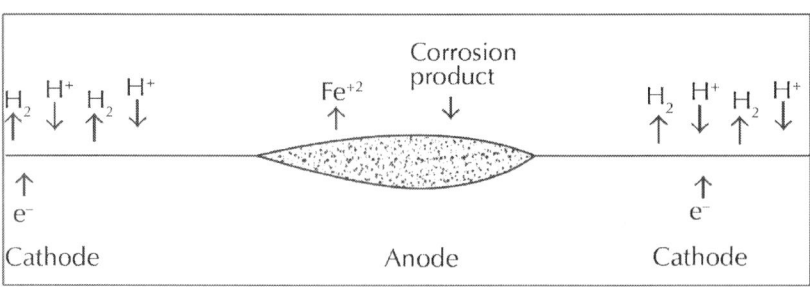

Figure 4: Acidic Condition of a Corrosion process.

The cathodic reaction equation is shown in (1), while the anodic equation is shown in (4).

In a deoxygenated solution, the hydrogen reaction combines with the others to yield the net corrosion reaction shown in (7):

$$Fe + 2H_2O \longrightarrow Fe(OH)_2 + H_2(g) \qquad (7)$$

In oxygenated aqueous systems, the oxygen reduction leads to a slightly different net corrosion reaction as shown in (8):

$$Fe + H_2O + \frac{1}{2}O_2 \longrightarrow Fe(OH)_2 \qquad (8)$$

Whereas in deoxygenated solution, hydrogen is evolved, in oxygenated system, oxygen is consumed. The evolved hydrogen acts as a catalyst for the formation of magnetite (Fe_3O_4) in deoxygenated water. Experiment shows that hydroxide readily decomposes into magnetite in deoxygenated water above 100°C [28] as indicated in (9):

$$3Fe(OH)_2 \longrightarrow Fe_3O_4 + H_2(g) + 2H_2O \qquad (9)$$

The net corrosion reaction with the magnetite as the final product is shown in (10):

$$3Fe + 4H_2O \longrightarrow Fe_3O_4 + 4H_2(g) \qquad (10)$$

In oxygenated solution, the ferrous oxide (Fe^{2+}) does not immediately precipitate out since it rapidly oxidizes to ferric oxide (Fe^{3+}), as a result, insoluble iron hydroxide is formed which is converted to hematite as shown in (11):

$$Fe(OH)_2 + \frac{1}{2}H_2O + \frac{1}{4}O_2 \longrightarrow Fe(OH)_3 \qquad (11)$$

The ferric oxide (Fe^{3+}) is converted to magnetite according to (12)

$$2Fe(OH)_3 + Fe(OH)_2 \longrightarrow Fe_3O_4 + 4H_2O \qquad (12)$$

Cathodic and anodic sites could be built as a result of variation in environmental conditions, metallic microstructure variation, and variation in environmental concentration of oxygen at different points of a metal [26]. At the anodic sites, the dissolution of metallic ions in the electrolyte brings about the flow of electrons between the corroding anodes and non-corroding cathodes. The spontaneous nature of the corrosion however, depends on the rate of flow of these electrons.

Though establishing the tendency for corrosion is necessary, however, it is more important to determine the rate of corrosion. This is because a particular metal or alloy may be prone to corrosion in an environment but at a very low rate, in which it will not be a problem [26]. To understand the rate of corrosion however, requires the knowledge of the role of primary environment and metallurgical variables, underlying mechanism of corrosion, and synthesis of information to account for effects of the parameters.

MECHANISMS OF CORROSION IN OIL AND GAS PIPELINES

Fluid flowing from oil and gas pipelines has a combination of chemicals including CO_2, H_2S, organic acids, bacteria, sand, and water. These constituents are among the major causes of corrosion in pipeline. The CO_2 dissolves in the presence of water to form an acidic oxide which reacts with iron. This type of corrosion is referred to as sweet corrosion. This is responsible for most types of general corrosion in oil and gas pipeline. Sour corrosion occurs when H_2S in the excess of 100ppm is present in the oil and gas, causes corrosion in the pipeline, and predominantly causes pitting [26, 29].

CO_2 present in oil and gas will dissolve in water to produce carbonic acid (H_2CO_3) [23, 27]. This acid dissolves steel to produce iron carbonate and hydrogen as shown in (13). This reaction takes place at the cathode:Despite the weakness of carbonic acid it is extremely corrosive to carbon steel. The chemical reactions above form the iron carbonate films. Depending on the condition during the formation, these films can be protective or non-protective at the anode, iron dissolves as shown in (4). The presence of CO_2 acts as a catalyst increasing the hydrogen evolution thereby increasing the corrosion rate of carbon steel in aqueous solution [27]. The carbonic acid (H_2CO_3) either serves as an extra source of H^+ or is reduced directly according to (14) and (15):The dissolved iron concentration will increase until Fe^{2+} is the same as the precipitation rate of $FeCO_3$ [30]. When Fe^{2+} is released in the corrosion process, the double amount of bicarbonate forms according to (16):The pH increases until bicarbonate and carbonate becomes so high that solid $FeCO_3$ precipitates [30] as shown in equation (17):When all the ferrous ions produced by corrosion precipitates as iron carbonate ($FeCO_3$), the pH remains constant and the overall reaction becomes as the state in (13).

In order to control the rate of corrosion on the pipeline, there should be passivity. Passivity is the condition existing on a metal surface because of the presence of protective film. When protective film is formed on the metal surface, it forms a coat which hinders further corrosion action on the material [26, 31]. The structure of the passive film (magnetite) formed on low carbon steel oxidizes in high temperature and has two distinct layers on the steel. The inner layer is compact and adheres well to the steel and has uniform thickness. The outer layer is a porous mass of individual crystal that would flake off the steel in some place and very nonuniform in thickness (Figure 5). This protective film is removed from the surface of the pipeline through erosion, dissolution, and turbulence resulting in more corrosion. The possible mechanisms resulting in the removal of the protective film are a follows:(i)Dissolution or removal of protective layer by hydrodynamic shear stress occurs when the shear stress is greater than the bonding force between the film and the substrate. This is a function of a mechanical process of erosion caused by the multiphase flow regime in pipeline [19, 32].(ii)In distributed flow condition, local near wall density of turbulence helps to remove the protective film. This disruption to the mass transfer boundary layer results in an enhanced corrosion rate [32, 33].(iii)Dissolution of film which is controlled by mass transfer. Thus the breakaway velocity may reflect conditions where the dissolution rate of the film is greater than the growth rate of the film [34].

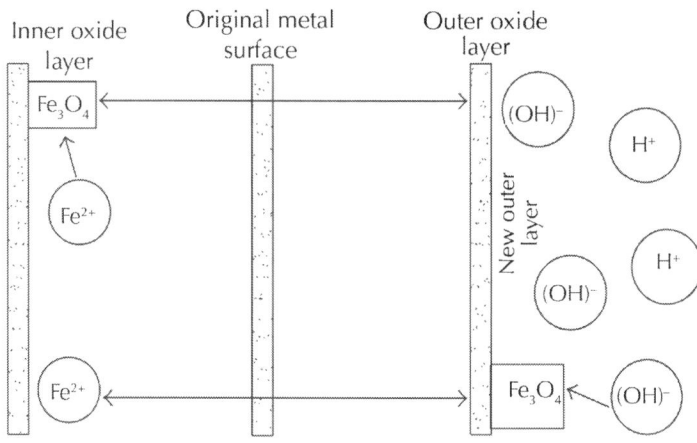

Figure 5: Schematic of magnetite double layer showing oxide formation locations.

The breakdown of protective film leads to the formation of localized corrosion that results in some of the major sources of corrosion failures like pitting, crevice, intergranular, and stress corrosion [12, 26, 35, 36]. The predominant breakdown processes are electrochemical and mechanical. Mechanical breakdown occurs when the protective film is ruptured as a result of stress or abrasive wear while, the electrochemical breakdown is a function of chemical reaction between the fluid constituent and the steel.

Types of Corrosion in Oil and Gas Pipeline

The primary chemical components that cause corrosion reaction to occur in pipeline are oxygen, acidic sulphur, and acidic chloride that dissolves in the water in the pipeline. The mechanism present in a given piping system varies according to the fluid composition, service location, geometry, temperature and so forth. In all cases of corrosion, the electrolyte must be present for the reaction to occur.

Internal Corrosion

Internal corrosion has become an increasing problem in most oil and gas pipelines as water cuts have increased and previously oil wet pipe surfaces have become water wet (providing the electrolyte for the corrosion cell) and as bacterial activities increases in the production system. Internal corrosion is the largest cause of pipeline failure in oil and gas industries [24] through different forms of corrosion like microbiologically influenced corrosion (MIC), erosion (flow enhanced) corrosion, under deposit (concentration cell) corrosion and so forth.

Erosion-corrosion

The erosion-corrosion mechanism increases corrosion reaction rate by continuously removing the passive layer of corrosion products from the wall of the pipe. The passive layer is a thin film of corrosion product that actually serves to stabilize the corrosion reaction and slow it down. As a result of the turbulence and high shear stress in the line, this passive layer can be removed causing the corrosion rate to increase [37]. The erosion-corrosion is always experienced where there is high turbulence flow regime with significantly higher rate of corrosion than just corrosion or erosion in pipeline [38]. In a multiphase flow regime with a fully developed turbulent flow, bubbles development and collapse have been attributed to changes in mass transfer coefficient and an eventual increase in CO_2 corrosion in pipeline [34].

Under Deposit Corrosion

The under deposit mechanism can increase the corrosion reaction rate by causing a localized chemical concentration which results in pitting of the metal surface under solid deposits. These deposits appear to be composed of a corrosion product matrix with entrapment of formation solids, sand, and iron sulphide. The rate of corrosion under this mechanism is significantly lower than erosion-corrosion mechanism.

Microbiologically Induced Corrosion (MIC)

This type of corrosion is caused by bacterial activities. The bacteria produce waste products like CO_2, H_2S and organic acids that corrode the pipes by increasing the toxicity of the flowing fluid in the pipeline. Some bacteria like sulphate removing bacteria (SRB) consume hydrogen that is a product in a standard corrosion reaction process. This activity causes the existing corrosion rate to increase in an attempt to reach reaction equilibrium by replacing the hydrogen consumed by bacteria. Bacteria also accumulate on the pipe walls, creating deposits and under deposit corrosion. MIC is recognized by the appearance of black slimy waste material or nodules on the pipe surface as well as pitting of the pipe wall underneath these deposits.

Pitting Corrosion

Pitting is classified as a localized attack that results in rapid penetration and removal of metal at small discrete area. The initiation of a pit occurs when electrochemical or chemical breakdown exposes a small local site on a metal surface to damaging species such as chloride ion. The site where pitting occurs is where there is an environmental variation in comparison to the entire metal surface. The combination of chlorine with H_2S results in localized pitting on steel [35]. This area of pitting which is usually the anode normally get highly degraded due to enormous electron transfer between the entire large area of the metal surface which is the cathode and small anode (the pitting site).

Crevice Corrosion

Crevice corrosion results when a portion of a metal surface is shielded in such a way that the shielded portion has limited access to the surrounding environment. Such surrounding environment contain, damaging corrosion species usually chloride ion. A typical example of crevice corrosion is the crevice found at the area between two metal surfaces in close contact with a gasket or another metal surface. The environment that eventually forms in the crevice is similar to that formed under the precipitated corrosion that covers a pit. An electrochemical corrosion cell is formed from the couple between the unshielded surface and the crevice interior exposed to an environment with a lower oxygen concentration compared with the surrounding medium. The concentration of being the anode of a corrosion cell and existing in an acidic, high-chloride environment where repassivation is difficult makes the crevice interior subject of corrosion attack.

Stress Corrosion Cracking (SCC)

Stress corrosion cracking (SCC) is a form of localized corrosion which produces cracks in metals by simultaneous action of a corrodent and tensile stress. It propagates over a range of velocities from $10^{-3} - 10$ mm/h depending upon the combination of alloy and environment involved. The geometry is such that if they grow to appropriate lengths, they may reach a critical size that results in a transition from the relatively slow crack growth rate associated with stress corrosion to fast crack propagation rates associated with purely mechanical failure. This transition happens when the stress intensity, which is a function of the geometry of the component including the crack size, reaches the fracture value for the material concerned. SCC in pipeline is a type of environmentally associated cracking (EAC). This is because the crack is caused by various factors combined with the environment surrounding the pipe. The most obvious identifying characteristic of SCC in pipeline is high pH of the surrounding environment, appearance of patches, or colonies of parallel cracks on the external of the pipe [39].

Top of the Line Corrosion (TLC)

This type of corrosion occurs due to the inability of corrosion inhibitors getting to the top of the pipeline (12 o'clock) thereby exposing it to corrodents. The inhibition effect is found to be predominant at the bottom of the line (6 o'clock), 9 o'clock and 3 o'clock where the flow of the oil or gas is taking place. This exposes the top of the line to concerted attack by the agents of corrosion with a resultant failure at some point. The primary factor that affects TLC is temperature which acts on the iron carbonate film formed. The combined effect of temperature fluctuation and condensation rate exposes the iron carbonate film to deterioration and consequently more corrosion. A study of the influence of gas flow rate on TLC shows that higher flow rate (which results in higher condensation rate) brings about more corrosion [40], while at a certain critical condensation rate, temperature and pH, TLC does not occur in gas pipelines [41]. The presence of acetic acid (HAc) has been found to enhance CO_2 TLC on carbon steel pipe, though at certain concentration level, HAc does not affect CO_2 TLC in carbon steel [42].

External Corrosion

External corrosion is caused by water penetrating the insulation system and is trapped between the insulation and the external pipe wall. The corrosion cell is fuelled by the continual supply of water and oxygen from the external sources. The main area where external corrosion is found is at the field applied weld insulation packs, but it can also be at any location where the galvanized insulation jacket has been punctured or torn. Weld pack insulations that are not well sealed allow water ingress making the weld packs to be wet. A fairly high temperature is needed to drive the corrosion mechanism, and the longer the mechanism has been active, the worst the damage will be. Therefore, the hottest and coldest lines in the field should have the highest likelihood for having an external corrosion problem.

CORROSION MANAGEMENT TECH-NIQUES

Corrosion management is that part of the overall management system, which is concerned with the development, implementation, review, and maintenance of corrosion policy [7]. The corrosion policy, however, is a framework on which decision concerning corrosion issue in an industrial setting is based. This framework provides basic measures for risk determination via development of absolute risk control measures through planning, implementation, and control strategies.

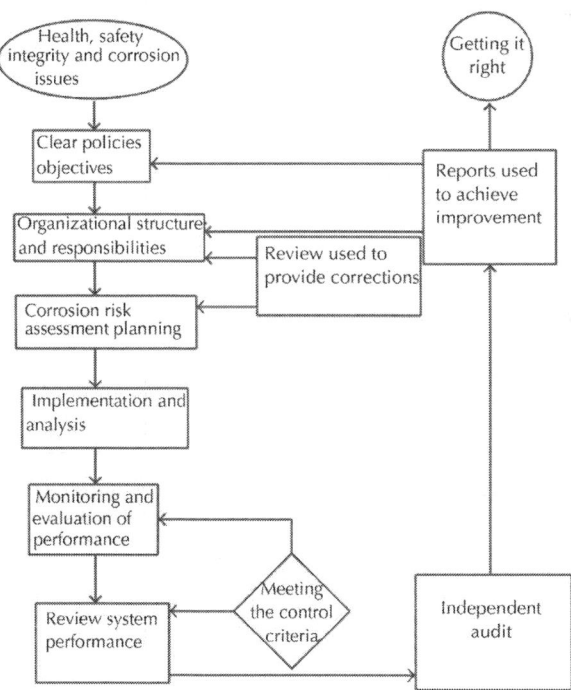

Figure 6: Corrossion management framework.

Corrosion management contributes to numerous benefits like statutory or corporate compliance with safety, health and environment policies, reduction in leaks, increased plant availability, reduced unplanned maintenance, and reduction in deferment costs [43]. To manage corrosion involves the utilization of a framework that will model the organization's policy through organizing, planning and implementing, measuring and reviewing, and auditing performance at all levels of execution as shown in Figure 6.

Corrosion Risk Assessment (CRA)

In planning for corrosion management, there is need for a formal identification of facilities that have the risk of degradation due to corrosion. The purpose of corrosion risk assessment is to rank facilities in order of their proneness to corrosion, identify options to remove, mitigate, or manage the risks. In order to manage corrosion risks, monitoring and inspection program will be incorporated in the overall activity schedule of an organization. The probability of failure is estimated based on the type of corrosion damage expected to occur on the component while the consequences of failure are measured against the impact of such a failure evaluated against a number of criteria. The criteria could include potential hazards to environment, risks associated with safety and integrity, or risk due to corrosion or inadequate corrosion mitigation procedure.

Typical of the risk-based procedure is the Failure Mode, Effect and Critical Analysis (FMECA) that ranks perceived risks in order of seriousness as shown in (18):where: failure Criticality is potential failures as examined in order to predict the severity of each failure effect in terms of safety, decreased performance, total loss of function and environmental hazards. Failure effect is potential failures assessed to determine the probable effect on process performance and the effects of the components on each other. Failure mode is the anticipated operational conditions used to identify most probable failure mode, the damage mechanism and likely locations. Corrosion risk is the product of the probabilities of a corrosion-related failure and the consequences of such a failure [44]. The risk analysis of a pipeline is a measure of the probability of failure. The acceptable annual failure probability is dependent on the safety class [45] as shown in Table 1.

Table 1: Safety class and target annual failure probability

Safety class	Annual failure probability
High	$<10^{-3}$
Medium	$<10^{-4}$
Low	$<10^{-5}$

Corrosion risk assessment can be carried out on a group of components which are constructed from the same material and subject to the same operating condition or an individual component. In oil and gas pipelines, the risk is analysed as either external or internal corrosion threat or environmental and operational threat. The remaining life of the pipeline is estimated against some established operational standards, while the rate of corrosion is correlated with the operating parameters of the oil and gas like CO_2, H_2S, temperature, pressure, flow rate, water cut and so forth. For effective corrosion assessment, the information concerning the operating condition of a facility will be maintained throughout the life cycle. The information is useful in formulating a corrosion risk assessment model that will be validated and modified with new assumptions overtime. For a non stable process condition, detailed re-assessment will be required at least annually but a stable process with good historical data trend will need revalidation less frequently [7].

Risk Based Inspection (RBI)

In managing oil and gas pipelines against corrosion, RBI technique is used to develop an optimum plan for the execution of the inspection activities. RBI uses findings from corrosion risk assessment (CRA) or other risk analysis to plan physical inspection procedures. A risk-based approach to inspection planning will ensure that risk is reduced to as low as reasonably practicable. It will also optimize inspection schedule, focus effort on the most critical area, and identify the most appropriate methods of inspection [46]. Planning a risk-based analysis involves listing activities, task and other elements of a project, identifying the technical risks, develop a risk ranking factor scale for each activity, document results and identify potential risks reduction actions for evaluation by personnel [47].

Corrosion Monitoring

Corrosion inspection and monitoring are key activities in ensuring, pipelines integrity are maintained and corrosion mitigated [48]. The choice of corrosion control measure is a function of fluid composition, pressure, temperature, aqueous fluid corrosivity, facility, and technical culture inherent in an establishment. In monitoring and inspection of pipelines, data are collected to enhance corrosion control by way of predicting the remaining life and the suggestion of possible mitigation measures that will help to enhance serviceability will largely depend on the experience of the personnel. A thorough practice for corrosion management involves the monitoring of corrosion risks through proactive and reactive monitoring techniques. In management of pipeline corrosion in oil and gas industries, proactive technique which involves determination of the corrosion standpoint prior to failure is utilized. This involves in-line and on-line monitoring system. In this system, data which could enhance the knowledge of the rate of corrosion degradation are collected and steps are taken to prevent failure. In-line system cover the installation of devices directly into the pipeline like corrosion coupons, biostuds and so forth. These need to be extracted for analysis periodically. On-line monitoring techniques include deployment of corrosion monitoring devices either directly into the process or fixed permanently to the facility. These include electrical resistance (ER) probes, linear polarization resistance (LPR) probes, fixed ultrasonic (UT) probes, acoustic emission and so forth.

Whereas some corrosion monitoring techniques can be used for continuous monitoring, others are used for periodic monitoring.

Corrosion monitoring techniques can either be direct or indirect parameter measure. This is summarized in Table 2.

Table 2: Summary of corrosion monitoring techniques

Direct method	Indirect method
Non-destructive inspection (NDI)	Biological counts
Material test coupons	Hydrogen probes
Electrical resistance (ER) probes	pH probes
Linear polarization resistance (LPR)	Specific ions
Elector-chemical impedance spectroscopy (EIS)	Temperature
Electro-chemical noise (EN)	conductivity
Galvanic current (GC)	Electrical potential monitor

Corrosion Mitigation Strategies

After corrosion risk assessment and data collection and analysis are completed, there is need for corrective action on the facility; this depends on the level of the deterioration experienced by facility. The approaches available for mitigating corrosion in pipeline includes, coating surfaces to act as a barrier or perhaps provide sacrificial protection, the addition of chemical specie to the environment to limit corrosion, alteration of alloy chemistry to make it more resistance to corrosion and utilization of alternative material [24].

Effective corrosion mitigation involves a good approach to assessment linked to inspection monitoring during initial design and re-evaluation of pipeline with respect to the selection of inhibitors. The summary of inhibitor selection for carbon steel pipeline at different risk categories is shown in Table 4.

Corrosion can be prevented or controlled by understanding the principle underlying corrosion process. This understanding has been the basis for the development of a number of corrosion prevention measures. The basic corrosion control measures are based on electrochemical driving force as shown in Pourbaix diagram in Figure 7. Table 3 shows the different pipeline corrosion mitigation strategies.

Table 3: Shows the different pipeline corrosion mitigation strategies

Mitigation strategy	Option	Remarks
Appropriate materials	Use of corrosion resistant alloys, non-metallic materials like Reinforced composite, thermoplastic-lined and polyethylene pipelines. Consider use of internally coated carbon steel pipeline systems (i.e., nylon or epoxy coated) with an engineered joining system.	(i) Non-metallic materials may be used as a liner or a free standing pipeline depending on the service conditions. (ii) Selection of appropriate material at construction and major refurbishment stage is necessary.
Chemical treatment	Corrosion inhibitors, biocides, oxygen scavengers, gas blanketing, vacuum deaeration	(i) The presence of small amounts of oxygen (parts per billion) or bacteria will accelerate corrosion. (ii) Provides a barrier between corrosive elements and the pipe surface
Coating and lining	Organic Coatings, metallic coatings, lining, cladding	Useful for internal and external corrosion prevention
Cathodic protection	Sacrificial anodes, impressed current systems, hybrid system	Need ability to monitor performance on-line.
Process control	Identify key parameters: pH, temperature, pressure, Flow rate, water chemistry, pH, chlorides, dissolved metals, bacteria, suspended solids, chlorine, oxygen, and chemical residuals	(i) Changes in operating conditions will influence the corrosion potential. Production information can be used to assess corrosion susceptibility based on fluid velocity and corrosivity (ii) Trends in dissolved metal concentration (i.e., Fe, Mn) can indicate changes in corrosion activity
Design detailing	Ensure ease of access and replacement:(i) Install valves that allow for effective isolation of pipeline segments from the rest of the system (ii) Install binds for effective isolation of in-active pipeline segments	Allows the effective suspension and discontinuation of pipeline segments:(i) Removes potential "deadlegs" from the gathering system (ii) Develop shut-in guidelines for the timing of required steps to isolate and lay up pipelines in each system

Table 4: Corrosion inhibitor risk categories

Risk category	Max inhibitor availability	Max expected un-inhibited corrosion rate (mm/yr)	comments	Proposed category name
1	0%	0.4	Benign Fluid, corrosion inhibitor use not anticipated. Predicted metal loss accommodated by corrosion allowance	Benign
2	50%	0.7	Corrosion inhibitor probably required but with expected corrosion rates there will time be time to review the need for inhibition based on inspection data.	Low
3	90%	3	Corrosion inhibition required for majority of field life but inhibitor facilities need not be available from day one.	Medium
4	95%	6	High reliance on inhibition for operational life time. Inhibitor facilities most be available from day one to ensure success	High
5	>95%	>6	Carbon steel and inhibition is unlikely to provide integrity for full field life. Select corrosion resistant material or plan for repair and replacement	Unacceptable

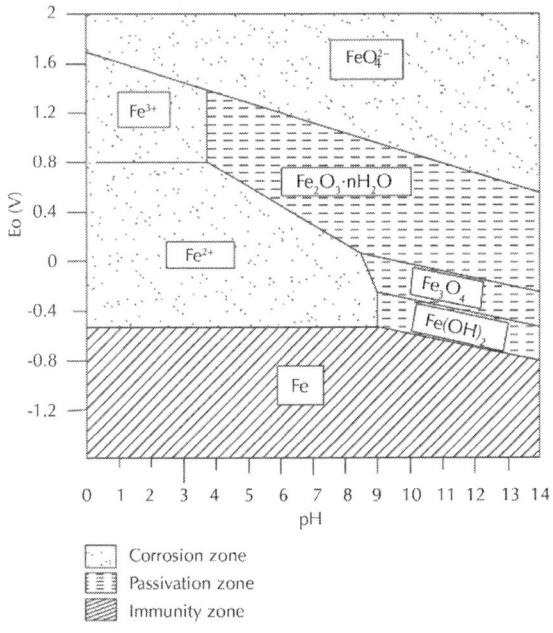

Figure 7: Pourbaix diagram of iron corrosion, passivity and immunity.

CONCLUSIONS

The prevalence of corrosion in oil and gas industry has resulted in enormous investment in technology to help combat impacts like loss of containment, leakages, death of personnel an environmental pollution. To this end, new oil and gas fields are developed using experiences generated from previous fields with similar characteristics. Efforts of design personnel at ensuring that the carbon steel materials are operating in the environment of immunity or passivity as shown in Pourbaix diagram (Figure 7) are yielding results via introduction of high corrosion resistant carbon alloys for pipelines. In other instances, specialized corrosion resistant materials have been used for lining the pipelines while reinforced composite and PVC materials have been utilized as alternative material for pipeline construction. Advancements in inspection and monitoring techniques are also aiding corrosion experts in decision concerning the "when" and "how" pipelines are managed in a bid to optimize performance and cost. The proliferation of different empirical, statistical, and mechanistic prediction models for corrosion prediction is aiding personnel in managing the integrity of the pipelines through different mitigation strategies.

Finally, if pipeline corrosion which is a major contributor to nonproductive time (NPT) in oil and gas production will be reduced to the barest minimum, a corrosion management policy with a well-defined structure that includes responsibilities, reporting routes, practices, procedures, and resources has to be strictly followed in the oil and gas industries. The effectiveness of the policy will therefore depend on the willing of the leadership and commitment of other personnel at all ranks.

REFERENCES

1. R. Nicholson, J. Feblowitz, C. Madden, and R. Bigliani, "The Role of Predictive Analytics in Asset Optimization for the Oil and Gas Industry-White Paper," 2010, http://www.tessella. com/wp-content/uploads/2008/02/IDCWP31SA4Web.pdf.

2. J. Neelamkavil, "A review of Existing Tools and Their Applicability to Maintenance Management. Report # RR-285,"http://pdf.aminer.org/000/274/575/a_decision_support_ system_for_transmission_facility_maintenance.pdf.

3. Oil and Gas Enhanced Production Services Industry to 2016—Enhanced Oil Recovery (EOR) Driving E&P Activity in Depleting Hydrocarbon Reservoirs, http://www.reportlinker. com/p0845623/Oil-and-Gas-Enhanced-Production-Services-Industry-to-Enhanced-Oil-Recovery-EOR-Driving-E-P-Activity-in-Depleting-Hydrocarbon-Reservoirs.html.

4. Control of Major Accident Hazards, "Ageing Plant Operational Delivery Guide,"http://www.hse.gov.uk/comah/guidance/ageing-plant-core.pdf.

5. P. Horrocks, D. Mansfield, K. Parker, J. Thomson, T. Atkinson, and J. Worsley, "Managing Ageing Plant," http://www.hse.

gov.uk/research/rrpdf/rr823-summary-guide.pdf.

6.   "Cost of Corrosion to Exceed $1 Trillion in the United States in 2012—G2MT Labs -The Future of Materials Condition Assessment," http://www.g2mtlabs.com/2011/06/nace-cost-of-corrosion-study-update/.

7.   Review of Corrosion Management for Offshore Oil and Gas Processing, HSE OffshoreTechnology Report 2001/044, 2001.

8.   Corrosion in the Oil Industry(Oilfield review) Schlumberger,http://www.slb.com/resources/publications/industry_articles/oilfield_review/1994/or19940401_corrosion.aspx.

9.   B. Khajota, D. Sormaz, and S. Nesic, "Case-based reasoning model of CO_2 corrosion based on field data," CORROSION, 2007, paper no. 07553.

10.   K. U. Raju, "Successful scale mitigation strategies in Saudi Arabian oil fields," in International Symposium on Oilfield Chemistry, The Woodlands, Tex, USA, April 2009, paper no. 121679.

11.   A. Darwin, K. Annadorai, and K. Heidersbach, "Prevention of corrosion in carbon steel pipeline containing hydrostatic water- an overview," in CORROSION, March 2010, paper no. 10401.

12.   J. Wen, T. Gu, and S. Nesic, "Investigation of the effects of fluid flow on SRB biofilm," in CORROSION, 2007, Paper no. 07516.

13.   B. Hedges, H. J. Chen, T. H. Bieri, and K. Sprague, "A review of monitoring and inspection technique for CO_2 and H_2S corrosion in oil and gas production facilities: location, location, location," inCORROSION, 2006, paper no. 06120.

14.   M. Singer, B. Brown, A. Camacho, and S. Neši , "Combined effect of carbon dioxide, hydrogen sulfide, and acetic acid on bottom-of-the-line corrosion," Corrosion, vol. 67, no. 1, 2011.

15.   K. L. J. Lee and S. Nesic, "EIS investigation of CO_2/H_2S corrosion," in CORROSION, April 2004, paper no. 04728.

16. K. D. Ralston and N. Birbilis, "Effect of grain size on corrosion: a review," Corrosion, vol. 66, no. 7, pp. 0750051–07500513, 2010.

17. Y. Song, A. Palencsár, G. Svenningsen, J. Kvarekvål, and T. Hemmingsen, "Effect of O_2 and temperature on sour corrosion," Corrosion, vol. 68, no. 7, pp. 662–671, 2012.

18. A. Kale, B. H. Thacker, N. Sridhar, and C. J. Waldhart, "A probabilistic model for internal corrosion of gas pipelines," in Proceedings of the 5th Biennial International Pipeline Conference (IPC '04), pp. 2437–2445, Calgary, Canada, October 2004.

19. S. Nesic, J. Cai, and K.-L. J. Lee, "A multiphase flow and internal corrosion prediction model for mild steel pipeline," in CORROSION, 2005, Paper no. 05556.

20. W. Sun and S. Nesic, "A mechanistic model of H_2S corrosion of mild steel," in CORROSION, 2007, paper no. 07655.

21. X. Hu, V. D. Souza, A. Neville, and J. Well, "Prediction of erosion-corrosion in oil and gas- a systematic approach," in CORROSION, 2008, paper no. 08540.

22. X. Tang, C. Li, F. Ayello, J. Cai, and S. Nesic, "Effects of oil type on phase wetting transition and corrosion in oil-water flow," in CORROSION, NACE International, 2007, Paper no. 017170.

23. Y. Xian and S. Nesic, "A stochastic prediction model of localized CO_2 corrosion," in CORROSION, 2005, paper no. 05057.

24. CAPP, "Best Management Practices: Mitigation of Internal Corrosion in Oil Effluent Pipeline Systems," 2009, http://www.capp.ca/getdoc.aspx?DocId=155641&DT=PDF.

25. B. A. Shaw and R. G. Kelly, "What is corrosion?" Electrochemical Society Interface, vol. 15, no. 1, pp. 24–26, 2006.

26. J. Kruger, "Electrochemistry of Corrosion," 2001, http://electrochem.cwru.edu/encycl/art-c02-corrosion.htm.

27. V. Fajardo, C. Canto, B. Brown, and S. Nesic, "Effect of organic acids in CO_2 corrosion," in Proceedings of the NACE International Conference and Exposition CORROSION, 2007, paper no. 07319.

28. P. S. Joshi, G. Venkateswaran, K. S. Venkateswarlu, and K. A. Rao, "Stimulated decomposition of $Fe(OH)_2$ in the presence of AVT chemicals and metallic surfaces—relevance to low-temperature feedwater line corrosion," CORROSION, vol. 49, no. 4, pp. 300–309, 1993.

29. R. Nyborg, "Controlling internal corrosion in oil and gas pipeline," Business Briefing-Exploration & Production: The Oil & Gas Review, no. 2, pp. 70–74, 2005

30. A. Dugstad, E. Gulbrandsen, M. Seiersten, J. Kvarekval, and R. Nyborg, "Corrosion testing in multiphase flow, challenges and limitations," in CORROSION, 2006, paper no 06598.

31. R. N. Kig, "A review of fatigue crack growth in air and seawater," Offshore Technology Report OTH96 511, HSE, 1996.

32. A. Keating and S. Nesic, "Prediction of two-phase erosion-corrosion in bends," in Proceedings of the 2nd International Conference on CFD in Minerals and Processes Industries CSIRO, Melbourne, Australia, December 1999.

33. S. Nesic and J. Postlethwaite, "Relationship between the structure of disturbed flow and erosion-corrosion," Corrosion, vol. 46, no. 11, pp. 874–880, 1990.

34. H. Wang, W. Paul Jepson, J.-Y. Cai, and M. Gopal, "Effect of bubbles on mass transfer in multiphase flow," in CORROSION, 2000, paper no. 00050.

35. H. Fang, B. Brown, and S. Nešiæ, "Effects of sodium chloride concentration on mild steel corrosion in slightly sour environments," in CORROSION, vol. 67, no. 1, January 2011.

36. E. Mysara Mohyaldinn, N. Elkhatib, and C. Mokhtar Ismail, "A computational tool for erosion/corrosion prediction in Oil/ Gas production facilities," in Proceedings of 3rd International

Conference on Solid State Science & Technology (ICSSST '10), Kuching, Malaysia, December 2010.

37. Sh. Hassani, K. P. Roberts, S. A. Shirazi, J. R. Shadley, E. F. Rybicki, and C. Joia, "Flow loop study of NaCl concentration effect on erosion, corrosion, and erosion-corrosion of carbon steel in CO_2-saturated systems," in CORROSION, vol. 68, no. 2, February 2012.

38. A. A. Sami and A. A. Mohammed, "Study synergy effect on erosion-corrosion in oil and gas pipelines,"Engineering and Technology, vol. 26, no. 9, 2008.

39. M. Baker Jr., "Stress Corrosion Cracking Study," 2004,http://www.polyguardproducts.com/products/pipeline/TechReference/SCC_Report-Final_Report_with_Database.pdf.

40. S. Olsen and A. Dugstad, "Corrosion under dewing conditions," in CORROSION, 1991, paper no. 472.

41. F. Vista and K. Alam, "Semi-empirical model for prediction of top-of-the-line corrosion risk," inCORROSION, 2002, paper no. 02245.

42. C. Mendex, M. Singer, A. Camacho, S. Hernndez, and S. Nesic, "Effect of acetic acid pH and MEG on CO_2 top of the line corrosion," in CORROSION, 2005, paper no. 05278.

43. D. Storey, "A Service Company's Experience with Pipeline Integrity Management," 2004,http://www.roseninspection.net/MA/papers/2004-11_PipelineIntegrityManagement.pdf.

44. P. O. Gartland and J. Roy, "Application of internal corrosion modelling in risk assessment of pipeline," in CORROSION, 2003, paper no. 03179.

45. Det Norske Veritas (DNV RPG 101), Recommended Practice DNV-RP-101: Corroded Pipelines, 2010.

46. Det Norske Veritas (DNV RPG 101), "Risk Based Inspection of Topsides Static Mechanical Equipment, 2001".

47. P. K. John and L. D. John, "Risk factor analysis—a new qualitative risk management tool," inProceedings of the

Project Management Institute Annual Seminar & Symposium, September 2000.

48. E. J. Carl, A. B. John, and G. T. Neil, "Improving plant reliability through corrosion monitoring," inProceedings of the Process Plant Reliability, Houston, Tex, USA, November 1995.

Chapter 3

Dust Explosion Prevention and Mitigation, Status and Developments in Basic Knowledge and in Practical Application

Rolf K. Eckhoff[1, 2]

[1]Department of Physics and Technology, University of Bergen, Allegaten 55, N-5007 Bergen, Norway
[2]Tyréns AB, 205 19 Malmö, Sweden

ABSTRACT

Right from the early days of the process industries, continuous efforts have been made to develop and improve measures for prevention and mitigation of dust explosions in these industries. Nevertheless this hazard continues to threaten industries that manufacture, use and/or handle powders and dusts of a wide range of combustible materials. To improve methods for predicting explosion development in real industrial plant has been one major challenge. Hence, during the last years comprehensive numerical simulation codes, for addressing this problem, have been developed. Progress has also been made in other areas, for example, ignition source prevention. The importance of adopting inherently safer process design, by building on firm knowledge in powder science and technology, and of systematic education/training of personnel, is also emphasized.

INTRODUCTION

Table 1 gives an overview of the most important methods currently used for preventing and mitigating dust explosions in the process industries. In dust explosion prevention and mitigation, as in many other challenges encountered by the process industries, there is an inevitable conflict between the short-term needs of the users of knowledge and technology and the long-term strive by researchers for the "perfect" solution. Industry will always need practicable tools and means that can be implemented more or less immediately. On the other hand, however, industrial pragmatism must not block the constant strive for better solutions based on improved basic understanding of the phenomena involved. A main aim of the present paper is to elucidate how fundamental research can promote further development of the practical means for preventing and mitigating dust explosions in industry that are listed in Table 1.

Table 1: Schematic overview of means of preventing and mitigating dust explosions

Explosion prevention		Explosion mitigation
Preventing explosive dust clouds	Preventing ignition sources	
Inerting of dust clouds by N_2+CO_2 and rare gases	Smouldering combustion in dust, dust fires	Explosion-pressure resistant construction
Intrinsic inerting of dust cloud by combustion gases	Other types of open flames (e.g., hot work)	Explosion isolation (sectioning)
Inerting of dust cloud by adding inert dust	Hot surfaces (electrically or mechanically heated)	Explosion venting
Keeping dust conc. outside explosive range	Heat from mechanical impact (metal sparks and hot-spots)	Automatic explosion suppression
Inherently safer process design	Electric sparks and arcs and electrostatic discharges	Partial inerting of dust cloud by inert gas
		Good housekeeping (dust removal/cleaning)

IN-DEPTH KNOWLEDGE—A POWERFUL AND ESSENTIAL TOOL IN ASSESSING AND CONTROLLING DUST EXPLOSION HAZARDS IN PRACTICE

Over the last 20 years, there has been a gradual shift in approach in dust explosion prevention and mitigation, from simple schematic design methods, toward more sophisticated ones opening up for increased flexibility and tailoring. During the same time period the appreciation of the benefits to be harvested from cross-fertilization between systematic research and practical applications has been growing. Advanced numerical models are starting to play an increasingly important role in solving practical design problems. The development of such models requires detailed experimental and theoretical studies of the relevant physical and chemical aspects. Table 2 summarizes some fundamental research topics that are essential for further development of the preventive and mitigatory methods and design tools that are indicated in Table 1. For example, basic understanding of flame propagation processes in dust clouds is the key to adequate design of practical mitigatory measures such as systems for dust explosion venting, suppression, and isolation. In Section3 some work on basic aspects of dust flames will be reviewed.

Table 2: Fundamental aspects addressed in dust explosion research

Dust cloud formation processes	Dust cloud ignition processes	Flame propagation processes in dust clouds	Blast waves generated by burning dust clouds
Inter-particle forces in dust deposits (cohesion)	General theories for ignition of single particles and clouds	Single-particle ignition and combustion in hot oxidizer gas	Blast wave properties as a function of properties of burning dust clouds
Entrainment of particles from dust deposits by shock waves passing across the deposit surface	Ignition by smouldering combustion in dust layers/deposits	Laminar and turbulent flames in dust clouds	Effects of blast waves on humans and mechanical structures
	Ignition by hot surfaces		

Entrainment of particles from dust deposits by turbulent gas flows Transport of dust particles in turbulent gas flows	Ignition by flying burning metal particles	Mechanisms of heat transfer (conduction, convection, radiation)	Ability of blast waves from dust explosions to transform dust layers into explosive dust clouds (coupled to first column of table)
	Ignition by electric sparks and arcs		
	Ignition by electrostatic discharges	Limit conditions for flame propagation in dust clouds (particle properties, dust conc., oxygen conc., geometry).	
	Ignition by hot gas jets		
	Ignition by shock waves		
Measurement and characterization of turbulence in dust clouds	Ignition by hot-spots from focused light beams	Acceleration of flames in dust clouds by turbulence mechanisms	
Measurement and characterization of spatial distribution of particles in dust clouds	Influences on dust cloud ignition sensitivity of cloud properties (composition, size, shape of particles, dust concentration, composition, turbulence, temperature and pressure of gas phase)	Detonation phenomena in dust clouds	

FLAME PROPAGATION IN DUST CLOUDS

Differences between Premixed Gases and Dust Clouds

In dust clouds, as opposed to premixed gases, inertial forces can produce fuel concentration gradients (displacement of particles in relation to gas phase). Furthermore, thermal radiation may contribute significantly to the heat transfer from the flame to the unburnt cloud, depending on the type of particle material (e.g., light metals). More work is needed to explore the role of thermal radiation in the development and course of dust explosions. Some papers discussing various central issues are those by Lee et al. [1] and Wolanski [2]. Much research work has been done on various aspects of combustion of liquid sprays and mists [3], which is in part also relevant even in the context of dust explosions.

Ignition and Combustion of Single Particles

Through the years a substantial amount of work has been conducted on various aspects of the ignition and combustion of single particles. A comprehensive review is given by Eckhoff [4]. Two more recent examples will be given here. One is the general study by Frolov et al. [5] on the effect of transient heat transfer on ignition of solid particles; the other is the investigation by Fedorov and Shulgin [6] on the stability of the process of ignition of small magnesium particles.

Flames in Dust Clouds

It has often been assumed that the laminar burning velocity of a given dust cloud is a basic combustion property of the cloud, which is closely related also to the burning velocities at various defined levels of turbulence and hence to the flame propagation through that type of cloud at large. An excellent recent contribution to improve understanding of the nature of laminar dust flames was given by Dahoe et al. [7].

Adequate submodels of flame propagation in turbulent dust clouds are essential in comprehensive numerical codes for dust explosion propagation. In the case of gases, extensive experimental research programmes have been conducted to resolve basic flame acceleration mechanisms in obstructed geometries. Central contributors are Moen et al. [8], Hjertager et al. [9], and Bakke and van Wingerden [10]. The fundamental studies of Rzal-Rebière and Veyssière [11] provide significant insight in possible differences between turbulent combustion of premixed gases and dust clouds. They investigated the interaction of a laminar maize starch/air flame with an obstacle, namely, a sphere, a disk, or an annulus.

One very interesting possibility would be to perform dust explosion experiments in large-scale experimental facilities used in previous extensive gas explosion experiments, for example, in the experiments by Moen et al. [8] on the influence of turbulence-generating baffles. By repeating these experiments with explosive dust clouds, and comparing the results with those found previously for gas, both important similarities and important discrepancies between turbulent dust and gas explosions could be disclosed.

Significant differences between combustion of premixed gases and dust clouds also exist on the microscopic scale. For example, the basic microscopic turbulence mechanisms that promote the combustion process must be identified. The results of Mitgau [12] and Mitgau et al. [13] indicate that more efficient replacement of gaseous reaction products by fresh air round each particle can be a strong basic turbulent combustion enhancement mechanism.

Cashdollar and Zlochower [14] measured flame temperatures and maximum explosions pressures in dust explosions with powders of a wide range of metals and sulphur, boron, and carbon. These data are useful in developing numerical models of dust flames.

Comprehensive Mathematical Models for Turbulent Flame Propagation in Dust Clouds

Kjäldman [15] was one of the pioneers in applying computational fluid dynamics (CFD) to turbulent dust explosion propagation. Subsequent contributions were made by Rose et al. [16], Smirnov et al. [17], Bielert and Sichel [18], Wörsdörfer et al. [19], Korobeinikov et al. [20], Zhong et al. [21], and Kosinski et al. [22]. Di Benedetto and Russo [23] presented a thermokinetic model of dust explosion propagation for natural and synthetic organic dusts. The model was based on the assumption that the devolatilization/pyrolysis step is very fast, and that the subsequent gas phase explosion is controlling the explosion rate. In developing a comprehensive numerical code for dust explosion simulation, corresponding existing codes for gas explosion simulation constitute a logical starting point. The comprehensive FLame ACceleration Simulator (FLACS) code, originally developed by Hjertager et al. [9] is currently being used as a basis for developing the corresponding dust explosion code Dust Explosion Simulation Code (DESC). Van Wingerden et al. [24], Arntzen et al. [25], Hansen et al. [26], and Siwek et al. [27] presented dust explosion simulations using preliminary versions of the DESC code. Skjold et al. [28–33] and Skjold [34], using improved versions of the same code, presented results from extensive simulations of dust explosion experiments performed in full scale process equipment, including a silo of 236 m^3.

Most probably this type of comprehensive numerical computer simulation code will become a future tool for predicting the course of complex dust explosion scenarios encountered in the process industries, for example, explosions propagating through a series of consecutive process units connected with ducts.

GENERATION OF EXPLOSIVE DUST CLOUDS IN PROCESS PLANT AND MEANS FOR THEIR PREVENTION

A Historical Perspective

Nearly 130 years ago Professor Weber, one of the pioneers of dust explosion research, stressed the importance of accounting for dust cohesion and dust dispersibility when considering the possibility of generation of explosive dust clouds. In his excellent paper on the ignitability and explosibility of wheat flour, Weber [35] emphasized that the cohesion of the flour, which is caused by interparticle adhesion, has a strong influence on the ability of the flour to disperse into explosive dust clouds. Weber suggested that two large dust explosion disasters, one in Szczecin (Stettin) and one in München, were mainly due to the high dispersibility of the flour. He also demonstrated, using simple but convincing laboratory experiments, that the dispersibility, or dustability, of wheat flour increased as its moisture content decreased. A global definition of dust dispersibility is given in Eckhoff [4, Chapter 3].

Generation of Primary Dust Clouds inside Process Equipment

In order for an explosive dust cloud to be formed from a layer/deposit, the layer/deposit must be exposed to a process that suspends the particles in the air to the extent that the dust concentration drops into the explosive range. Most often such dispersion of dust to form explosive clouds takes place intentionally inside process equipment, for example, by handling and transportation in various process equipment (e.g., mills, dryers, mixers, bucket elevators and other conveyors, silos, filters, cyclones, and connecting ducts). It is foreseen that in a not-too-distant future comprehensive numerical codes will be available for predicting the dust cloud structures (spatial distributions of effective particle size, dust concentration, turbulence, and global flow) that will be generated in various practical scenarios in industry. Knowing this initial cloud structure is essential both for predicting the ignition sensitivity of the cloud with regard to various ignition sources and for predicting the course of development of the primary explosion that will result from ignition. Therefore, adequate information about initial dust cloud structures is essential for realistic assessment of the dust explosion risk in a process plant. However, the development of adequate numerical models of dust cloud structures is not far beyond its infancy, and information of practical use is scarce. The works of Hauert et al. [36] and Kosinski et al. [37] constitute valuable initial contributions. More recent contributions are by Kosinski and Hoffmann [38], Kosinski et al. [39], and Ilea et al. [40]. However, the problem addressed is very complex and more experimental, and theoretical work is needed.

Generation of Secondary Dust Clouds inside and/or Outside Process Equipment by Blast Waves from Primary Dust Explosions

The blast wave from a primary dust explosion can generate secondary explosive clouds ahead of the flame by entraining dust deposits and layers there. Lebecki et al. [41], being primarily concerned with coal mine explosions, investigated this process experimentally in a 100 m long gallery of cross-section 3m^2. Kauffman et al. [42] and Austin et al. [43] summarized their extensive research on blast-wave entrainment of dust layers in long tubes, whereas Boiko and Poplavski [44] studied the effect of the dust concentration in a dust cloud behind a shock wave, on the acceleration of the cloud. Data from this kind of work are essential in the development of comprehensive dust explosion codes. Klemens et al. [45] presented a mathematical model for simulating the process of entrainment of dust particles from a dust layer, by the gas flow behind a shock or a rarefaction wave passing across the layer. The above mentioned subsequent contributions by Kosinski and Hoffmann [38], Kosinski et al. [39], and Ilea et al. [40] are indeed relevant also in the present context. The mathematical models of Fedorov and Gosteev [46] and Fedorov and Fedorova [47] describing the initial stage of the entrainment of single dust particles from a dust layer by a gas flow passing across the layer are also important contributions.

Dust Dispersibility Tests

Various test methods have been proposed for evaluating the ease with which dust clouds can be produced from deposits and layers of powders/dusts [4, Chapter 7], [48–51].

Inherently Safer Process Design to Prevent/Limit Generation of Explosive Dust Clouds

Most commonly the dust explosion hazard is combated by adding preventive and mitigatory measures to an existing process. However, the technical measures adopted are often expensive, and safety procedures may fail.

Inherent safety is an alternative approach. It implies that the process itself be designed in such a way that no explosion hazard exists. Kletz [52], the "father" of the inherently-safer-process design concept, outlined its basic philosophy and recommended the use of it whenever feasible. In the context of preventing and mitigating dust explosions inherently safer process design could include use of production, treatment, transportation, and storage operations where dust cloud generation is kept at a minimum. One example is the use of mass flow silos and hoppers instead of funnel flow types. Eckhoff [53] emphasized the importance of knowing powder science and technology when striving for inherently safer process design in industries having a dust explosion hazard. Amyotte and Khan [54] proposed a framework for directing the concept of inherently safer process design specifically toward reducing the dust explosion hazard in industry. Recently Amyotte et al. [55] described in greater detail how the inherent safety principles of minimization, substitution, moderation, and simplification can be implemented in practice to prevent and mitigate accidental dust explosions in process plant. Hopefully such initiatives will promote further work in this important area.

Inerting by Adding Inert Gas

Explosive dust clouds can be made inert by mixing the air with an inert gas such as nitrogen or carbon dioxide to a level at which the dust cloud can no longer propagate a self-sustained flame. Some further insight has been gained during the last two decades. Wilén et al. [56] found that the limiting oxygen concentration (LOC) for inerting of biomass dust clouds increased with increasing initial pressure of the cloud in the range 5 to 18 bar. This is opposite to the decrease of LOC with increasing initial pressure found earlier for clouds of coal dust. As would be expected, Schwenzfeuer et al. [57] found that LOC for ignition of dust clouds by electrostatic discharges, or metal sparks from mechanical impact, was significantly higher than the conservative limit determined in standard tests, using a very strong pyrotechnical ignition source.

Whilst adding nitrogen to the air can prevent dust explosions, it may introduce a suffocation risk. However, it has been shown that addition of a few vol. % of CO_2 to the nitrogen/air mixture reduces the critical oxygen threshold for suffocation considerably . A gas mixture for inerting utilizing this effect was presented by Dansk Fire Eater A/S [58].

Keeping the Dust Concentration below LEL

In principle, keeping the concentration of dust in the cloud below the lower explosive limit (LEL) is a means of maintaining dust clouds nonexplosive. However, with a few exceptions the method has limited applicability in practice in the process industries. Mittal [59] discussed various mathematical models for calculating minimum explosive concentrations of dust clouds.

PREVENTING IGNITION SOURCES

Smouldering Layers, Deposits, and Nests

Can metal particle sparks from single accidental impacts initiate combustion in dust layers/deposits? Hesby [60] found that the number of sparks from single accidental impacts of steel objects is far too low to be able to cause ignition of the layers of a selection of organic dusts, including tobacco. Gummer and Lunn [61] found that, in general, smouldering nests were poor ignition sources for most dust clouds, whereas flaming nests caused ignition more readily. More work is needed to clarify both the conditions under which smouldering or flaming nests of various materials are generated in industrial plant and the circumstances under which such nests will ignite explosive clouds of various dusts.

Krause and Hensel [62] presented a numerical method by which nonsteady temperature fields in dust deposits can be computed. This enables numerical analysis of a number of practical cases of self-heating/self-ignition that cannot be analyzed using the classical thermal explosion theory of Frank-Kamenetzki. Krause and Schmidt [63] studied experimentally critical thermal conditions that may lead to initiation of smouldering processes, or to further development of such processes, once initiated.

Hot Surfaces

In the past, the minimum hot-surface temperature for ignition of a dust cloud has often been regarded as if it were a universal constant for a given cloud. Consequently, results from small-scale laboratory tests were often applied directly in design of large-scale industrial plant. However, minimum hot-surface ignition temperatures of dust clouds vary significantly with scale as well as with the geometry of the hot surface in relation to the dust cloud. There is a need for both a more differentiated basic understanding and a more differentiated testing approach. Development of numerical models for dynamic simulation of hot-surface ignition processes encountered in practice is foreseen.

Electric/Electrostatic Discharges between Two Metal Electrodes

Electric and electrostatic discharges between two metal electrodes can be generated in a number of ways, for example, in switches, by failures in electric circuits and by discharge of static electricity. The parameters influencing the minimum energy required for igniting a dust cloud by an electric spark include voltage and current characteristics across the spark gap, spark gap geometry, and electrode material as well as all the dust cloud parameters. The latter include particle material and particle size/shape distributions, dust moisture content, dust concentration, and the dynamic state of the dust cloud with respect to the spark gap. Minimum ignition energies (MIEs) of clouds of a given dust material decrease strongly with the fineness of the dust.

Eckhoff [64] discussed the influence of dust fineness on MIE of ferro-alloys dusts. In the past dust fineness was often specified just as a mass percentage finer than an arbitrary size, for example, 74 m or 63 m, without any specification of the distribution of particle sizes below these limits. This complicates the analysis of published experimental data, and more systematic research is needed to clarify the exact influence of particle size. In the case of metal alloys the most hazardous components may sometimes accumulate in the fine tail of the particle size distribution (e.g., Mg in MgFeSi), and special investigations are required.

Lorenz and Schiebler [65] presented the results from a comprehensive, detailed experimental and theoretical investigation of the energy transfer processes taking place during an electrostatic spark discharge. The temperature and pressure development in the spark channel during its formation and subsequent expansion were investigated. This also included cooling of the channel by thermal radiation. The dependence of the ability of a given discharged electrical energy to ignite a dust cloud on these basic physical spark characteristics was emphasized.

Randeberg and Eckhoff [66] investigated an alternative method for measuring MIEs of explosive dust clouds, which may be in better accordance with accidental electrostatic spark ignition in industrial plant. They used the transient dust cloud itself to initiate spark breakdown between a pair of electrodes preset at a high voltage somewhat below the breakdown voltage in dust-free air. The MIEs obtained were of the same order as those obtained using the conventional synchronized-spark methods. The lower spark energy limit for apparatuses commonly used for determining MIEs of dust clouds has been 1–3 mJ, but more recently Randeberg et al. [67] presented a new test method that in principle permits MIE determination for dust clouds, using synchronized sparks, down to the order of 0.1 mJ. However, as pointed out by Eckhoff et al. [68], due to the design of the spark generator used, the spark energies quoted by Randeberg et al. were in fact considerably smaller than the real energies in their experiments. Work is currently being conducted at the University of Bergen to eliminate this problem. Whether new MIE measurements using the improved spark discharge circuit will necessitate adjustment of the very low MIEs for some dust clouds reported by Randeberg and Eckhoff [69] and Eckhoff and Randeberg [70] remains to be seen. Recently Wu et al. [71], using a measurement system with a lower spark energy limit of 1 mJ, reported that clouds in air of a number of very fine titanium and iron powders could be readily ignited by a spark of 1 mJ energy, which means that their MIEs were in fact significantly lower than this value.

Baudry et al. [72], Nifuku et al. [73], and Marmo and Caravello [74] measured MIEs of clouds in air of various types of dust (aluminium with various contents of oxide, aluminium and magnesium dusts from shedding processes, and nylon fibres, resp.).

Electrostatic One-Electrode Discharges

With regard to the even more complex one-electrode types of electrostatic discharges (corona, brush, propagating brush, etc.), valuable experimental insight has been gained during the last years. For example, the issue of whether brush discharges can ignite dust clouds in air was revisited experimentally by Larsen et al. [75], who were able to ignite clouds of fine sulphur dust in oxygen-enriched air by true brush discharges. However, ignition in air only was never observed. Because of the very low MIE of the sulphur dust used, this indicates that ignition of even the most sensitive dust clouds by brush discharges in air is unlikely.

Glowing/Burning Particles

Ignition of dust clouds by small burning metal particles called impact sparks or metal particle sparks is a complex process. Such sparks are generated by single, fast impacts between solid materials, of which one is a metal. So far practically useful theories, describing both impact and ignition, do not seem to be within sight. Such theories must comprise several complex subprocesses. The first is the generation and initial heating of the metal particle by the impact. The second is the ignition of the flying hot particle and the subsequent burning process. The third is the heat transfer to the dust cloud, which ultimately determines whether ignition occurs or not.

Electrical Apparatus

The present situation internationally concerning standards for electrical apparatus for use in areas containing combustible dust is confusing, as discussed by Eckhoff [76, 77]. The International Electrotechnical Commission (IEC) has decided to base its development on the European Union "Atex" philosophy. However, the European Union Atex 94/9/EC Directive does not distinguish adequately between combustible dusts and combustible gases/vapors. This has given rise to undue alignment of a series of new IEC standards for electrical apparatus for combustible dusts with established standards for gases/vapors. The current European Union Atex 1999/92/EC Directive also lacks the required distinction between gases and dusts, which gives rise to problems with area classification.

Other Ignition Sources

Proust [78] determined experimentally the minimum laser beam power required for igniting dust clouds by the heat absorbed by a solid target heated by the laser beam. The variable parameters included the laser beam diameter, the duration of the irradiation, the target material (combustible/noncombustible), and the type of dust. Initiation of dust explosions by shock waves has been studied by several workers, including Wolanski [2] and Klemens et al. [79].

PROTECTIVE/MITIGATORY MEASURES

Full Confinement

The applicability of the concept is limited because of high equipment costs. However, the method is used in some special cases, for example, when the powder/dust is highly toxic, and completely reliable confinement is absolutely necessary. Whereas current experimental methods allow accurate prediction of maximum attainable explosion pressures in simple vessels with point source ignition, design of pressure resistant process equipment may not be straightforward. The use of "finite element" computation methods to achieve improved design seems to be increasing.

Explosion Isolation

The objective of explosion isolation is to prevent dust explosions from spreading from the primary explosion site to other process units, workrooms, and so forth. Basic understanding of flame propagation and pressure build-up in coupled process equipment ("interconnected vessels") is required for specification of performance criteria of various types of active and passive isolation equipment. Van Wingerden et al. [80] reported on dust explosion experiments in a system of two vented vessels connected by a duct. Holbrow et al. [81, 82] summarized the results from extensive similar experiments in the UK and presented coherent quantitative guidance for design of interconnected process equipment based on containment and explosion venting. Vogl and Radandt [83, 84] presented results from a comprehensive experimental program in Germany on propagation of dust explosions in interconnected process systems. Various passive and active techniques for interrupting explosions in pipelines have been developed, but there is room for further improvement.

Partial Inerting

This is a relatively new concept for mitigating dust explosions, which deserves some attention. The idea is that as the oxygen content in the atmosphere is reduced, there is a systematic decrease of both ignition sensitivity and combustion rate of the dust cloud. In many cases the explosion hazard may be reduced markedly by only a moderate reduction of the oxygen content. Glor and Schwenzfeuer [85] confirmed experimentally that even modest reductions of the oxygen content can increase the minimum ignition energies of dust clouds substantially. Devlikanov et al. [86] found that K_{st} was a linear function of the percentage of oxygen in the gas phase (mixture of nitrogen and oxygen). Conde Lázaro and García Torrent [87] carried out a series of partial inerting experiments at 12 bar initial pressure, in a demonstration pulverized coal power plant. Eckhoff [88] called for more extensive use of partial inerting in industrial dust explosion protection at large.

Explosion Venting

This is probably the most widely used method for mitigating dust explosions. In spite of extensive research and development, dust explosion venting remains a complex and in part controversial subject. The key issue is vent area sizing.

Tamanini and Valiulis [89] presented an improved version of the VDI (Germany) and NFPA (USA) guidelines for sizing of dust explosion vents. The improvement was achieved by systematizing the data in the context of a simplified physical model of the vented explosion. A similar contribution was made by Ural [90]. The new CEN [91] standard for design of dust explosion venting systems in principle opens up for a differentiated approach to vent sizing, which accounts for the variations in dust cloud structures encountered in practice in industry. In most practical cases this will result in more liberal vent area requirements than those of some previous rigorous standards.

Other aspects of explosion venting studied more recently include the influence of the inertia and specific design of the vent cover on the gas dynamics of the venting process. A further dimension of complexity is added to the venting problem if the initial pressure (and/or temperature) deviates from atmospheric. Results from venting of dust explosions in air at elevated initial pressure were reported by Siwek et al. [92].

In dust explosion venting, maintaining the integrity of the enclosure is not the only concern. Venting implies that both blast waves and flames are emitted into the surroundings, and this may present a hazard, depending on the size of the emitted flame and the magnitude of the blast wave. Several workers, including Forcier and Zalosh [93], Holbrow et al. [94] and Harmanny [95], investigated various aspects of this problem. Various methods have been developed for eliminating hazardous effects of flames from vent openings. Li et al. [96] and Emde and Penno [97] discussed further aspects of the Q-pipe for dust and flame-free venting. The influence of vent ducts on the maximum explosion pressure in the vented vessel has been studied experimentally by several workers including Ural [98] and Lunn [99]. Tamanini and Valiulis [100] presented a new theoretical approach for predicting the resultant reaction impulse acting on a process structure during a vented explosion.

Venting of industrial buildings requires special considerations. An overview was given by Crowhurst [101]. Höppner [102] discussed the design of dust explosion venting arrangements for rooms/buildings of volumes 5000m^3, with walls that can only withstand overpressures less than 0.2 bar. In case of a dust explosion, only part of such large volumes will be filled with explosive dust cloud.

Tamanini [103] summarized his valuable effort of correlating existing experimental dust explosion venting data by applying the classical method of dimensional analysis. It is regrettable that this important work was not included in the recent European Union guideline for design dust explosion venting arrangements, CEN [91], but it has been included in recent US standards (NFPA). Silvestrini et al. [104] developed correlations for flame speed and explosion overpressure for dust explosions inside industrial enclosures with the aim to provide a simple tool for sizing dust explosion vents.

However, in view of the different turbulence levels, degrees of dust dispersion, and distributions of dust concentrations encountered in industry, the need for a more complete differentiated approach to assessment of vent area requirements has gained general acceptance. As indicated by Skjold [34], the numerical code DESC may become a useful tool to meet this need.

Automatic Explosion Suppression

This active method for dust explosion mitigation is comparatively complex and expensive. It is therefore used when simpler and less expensive methods are inadequate. The method has been in use for many decades, and significant progress has been made during the last decade. For example, Moore [105] and Chatrathi and Going [106] evaluated the suitability of various suppressants, and Tyldesley [107] reported that superheated water can in some circumstances be an effective candidate. Moore and Siwek [108] summarized their extensive multiyear experimental work on suppression of dust explosions, whereas Chatrathi and Going [109] gave an overview of current technology and philosophy for implementing automatic explosion suppression systems in practice. The influence of elevated initial temperature of the explosive dust cloud on the efficacy of an automatic explosion suppression system was studied by Brehm [110].

The European standardization organization CEN [111] has produced a draft standard for design of explosion suppression systems, which seems to open up for greater flexibility than the traditional, mostly very conservative approach. Hence, if the turbulence level and/or degree of homogeneity of the cloud of a given dust in the actual process situation are lower than those produced by the rather conservative traditional standard VDI-method of dust cloud generation, this can be accounted for in the design of the suppression system.

Comprehensive numerical models of the complex explosion suppression process, based on computational fluid dynamics (CFD), are likely to be the future tool for design of optimal explosion suppression systems. Again the code DESC should be mentioned as a promising candidate. Morgan [112] assessed the suitability of commercially available CFD software for modeling the types of flows encountered in explosion suppression processes. Using results from his model simulations, he was able to design a novel suppressant injection nozzle, which was shown to be more effective than standard nozzles currently used.

Design of Process Equipment for Specific Internal Explosion Loads

This problem is also addressed in 6.1 above and is a central concern also when designing explosion venting systems (6.4) and systems for automatic explosion suppression (6.5). Harmanny [113] presented a new equation for predicting the duration of vented dust explosions in enclosures of volumes from 10 to 60m³. This is a useful tool for evaluating whether static pressure considerations or impulse considerations apply when predicting the response of the enclosure structure to the explosion load. Harmanny [114, 115] revisited the problem of assessing the structural response of a given process equipment and buildings to explosion loads. With regard to dust explosions in the process industries, he concluded that most often they are sufficiently slow for the load to be regarded as quasistatic. However, there are certain cases where dynamic effects play a significant role. Comprehensive finite-element-based computer codes for determining detailed stress/strain analyses of complex structures exposed to defined static and dynamic loads have been available for some time. It is foreseen that the use of such tools in assessing the explosion strength of complex process equipment will increase in the years to come. The concept of pressure-shock-resistant design should be developed further to facilitate cost effective equipment design. Li et al. [116] compared elastic and plastic structural responses of a simple mechanical structure determined experimentally with predictions from using a computational finite-element approach.

Preventing Secondary Explosions outside Process Equipment

This remains a most important issue in all efforts to fight the dust explosion hazard. Adequate housekeeping is an essential means of achieving this aim. However, there are still questions to be answered concerning the level of cleanliness required. More research is needed for assessment of the maximum acceptable mass of deposited dust per unit area of surface for preventing secondary dust flame propagation under various conditions. Cybulski et al. [117] showed that comparatively weak secondary dust explosions in short narrow tunnels in grain elevators can be extinguished by properly designed, actively triggered water barriers. They also showed that, under the conditions prevailing, the possibility of flame penetration into blind gallery branches was small. This kind of work may also be of relevance to the analysis of flame propagation in large industrial systems, for example, in grain storage and handling plants.

OTHER FACTORS INFLUENCING THE DUST EXPLOSION RISK

Explosion Risk Management

Barth [118] emphasized the importance of companies establishing systems for explosion risk management control to ensure effective, long-lasting explosion protection of process plant. Hesener et al.

[119], with reference to the pharmaceutical industry, underlined the need for having adequate systems for explosion risk management and control even in small and medium size plants. Van der Voort et al. [120] presented a quantitative risk assessment tool for the external safety of industrial plants with a dust explosion hazard.

Cost/Benefit in Dust Explosion Prevention and Mitigation

Alfert [121] addressed the bottom-line costs of various dust explosion protection systems on the market. Janssens [122] pointed out that the investments required to achieve proper prevention and control of the explosion hazard in a given plant are not necessarily excessive. By combining thorough knowledge of the processes to be protected, with knowledge of relevant ignition and flame propagation phenomena, and principles and technologies available for explosion control, good solutions can be obtained at an acceptable cost.

Education and Training

High safety levels in the process industries cannot be established once and for all by a single all-out effort. Deterioration results if the high level once attained is not actively secured by continuous maintenance and renewal. This applies both to technology and human factors. Education and training, from short practical training courses to in-depth long-term education, play a key role in the continuous maintenance and renewal process. Universities and colleges have responded to this challenge by establishing study courses on a wide range of process safety aspects. Relevant topics include reliability and risk analysis, the physics, chemistry, and technology of processes and hazards, and means of accident prevention and mitigation. Much emphasis has been put on methods of reliability and risk analysis, which are indeed very important. However, it is sometimes felt by the process industry itself that education in the "hard" aspects, that is, the physics, chemistry, and technology of processes and process hazards, has been somewhat left behind. This situation presents a special challenge to universities and colleges.

PERSPECTIVES FOR THE FUTURE

- The approaches taken in dust explosion prevention and mitigation in the process industries will become steadily less dogmatic and more tailored and differentiated in the years to come. Industry will strive for steadily more cost effective safety measures.

- Therefore, substantial progress is foreseen in mathematical CFD-based modeling of dust cloud generation and flame propagation processes in dust clouds. It is anticipated that such models will gradually replace conventional empirical equations and graphs as design tools for tailored systems for explosion isolation, explosion venting, and automatic explosion suppression and for evaluating consequences of secondary dust explosions. However, extensive experimental validation of the numerical models is absolutely necessary.

- Evaluation of potential ignition sources will also become more detailed and differentiated, in accordance with reality. Further development of mathematical models capable of predicting ignition of dust clouds and layers by self-heating/ smouldering, hot surfaces, various electrical and electrostatic sparks/discharges, metal sparks, and so forth, is foreseen.

- Combined protective solutions, for example, partial inerting in combination with venting, or venting combined with automatic suppression, are likely to be used to an increasing extent.

- The need for inherently, safer design of processes for production, treatment and handling of combustible powders/ dusts is expected to be become more and more accepted. To achieve this, knowledge and systematic use of **powder/** particle science and technology is a basic requirement.

- High-quality training/education, ranging from short courses of a few days to extensive university studies, will continue to be essential for minimizing the hazards in the process industries, including minimizing the risk of dust explosions.

REFERENCES

1. J. H. S. Lee, F. Zhang, and R. Knystautas, "Propagation mechanisms of combustion waves in dust-air mixtures," Powder Technology, vol. 71, no. 2, pp. 153–162, 1992. ·

2. P. Wolanski, Deflagration, Detonation and Combustion of Dust Mixtures, American Institute of Aeronautics and Astronautics, New York, NY, USA, 1990.

3. R. K. Eckhoff, "Generation, ignition, combustion and explosion of sprays and mists of flammable liquids in air. A Literature Survey," Tech. Rep. CMI–91–A25014, Christian Michelsen Institute, Fantoft, Norway, 1991.

4. R. K. Eckhoff, Dust Explosions in the Process Industries, Gulf Professional Publishing/Elsevier, Boston, Mass, USA, 3rd edition, 2003.

5. S. M. Frolov, K. A. Avdeev, and F. S. Frolov, "Effect of transient heat transfer on ignition of solid particles," Journal of Loss Prevention in the Process Industries, vol. 20, no. 4–6, pp. 310–316, 2007.

6. A. V. Fedorov and A. V. Shulgin, "About stability of the ignition process of small solid particle," Journal of Loss Prevention in the Process Industries, vol. 20, no. 4–6, pp. 317–321, 2007.

7. A. E. Dahoe, K. Hanjalic, and B. Scarlett, "Determination of the laminar burning velocity and the Markstein length of powder-air flames," Powder Technology, vol. 122, no. 2-3, pp. 222–238, 2002.

8. I. O. Moen, J. H. S. Lee, and B. H. Hjertager, "Pressure development due to turbulent flame propagation in large-scale methane—air explosions," Combustion and Flame, vol.

47, pp. 31–52, 1982.

9.   B. H. Hjertager, K. Fuhre, and M. Bjoerkhaug, "Gas explosion experiments in 1:33 and 1:5 scale offshore separator and compressor modules using stoichiometric homogeneous fuel/air mixtures," Journal of Loss Prevention in the Process Industries, vol. 1, no. 4, pp. 197–220, 1988.

10.   J. R. Bakke and K. van Wingerden, Guidance for Designing Offshore Modules Evolving from Gas Explosion Research, Society of Petroleum Engineers, Richardson, Tex, USA, 1992.

11.   F. Rzal-Rebière and B. Veyssière, "Propagation mechanisms of starch particles-air flames," inProceedings of the 6th International Colloquium on Dust Explosions, D. Xufan and P. Wolanski, Eds., pp. 186–200, Shenyang, China, August-September 1994.

12.   P. Mitgau, Einfluss der Turbulenzlänge und der Schwankungsgeschwindichkeit auf die Verbrennungs-geschwindigkeit von aerosolen, vol. 14, Max-Planck-Institut Für Strömungsforschung, Göttingen, Germany, 1996.

13.   P. Mitgau, H. Gg. Wagner, and R. Klemens, "Einfluss der Turbulenzlänge und der Schwankungsgeschwindichkeit auf die Flammengeschwindigkeit von Stäuben," in Feuerungstechnik, Kaleidoskop aus aktueller Forschung und Entwicklung. Geburtstag, Festschrift an Prof. Wolfgang Leuckel zu seinem 65, pp. 17–45, Engler-Bunte-Institut, Bereich Feuerungs-technik, Universität Karlsruhe (TH), Geburtstag, Germany, 1997.

14.   K. L. Cashdollar and I. A. Zlochower, "Explosion temperatures and pressures of metals and other elemental dust clouds," Journal of Loss Prevention in the Process Industries, vol. 20, no. 4–6, pp. 337–348, 2007.

15.   L. Kjäldman, "Numerical flow simulation of dust deflagrations," Powder Technology, vol. 73, no. 1, p. 100, 1992.

16.   M. Rose, P. Roth, S. M. Frolov, and M. G. Neuhaus, "Modelling of turbulent gas/particle combustion by a Lagrangian PDF method," Combustion Science and Technology, vol. 149, no.

1, pp. 95–113, 1999.

17. N. N. Smirnov, V. F. Nikitin, and J. C. Legros, "Ignition and combustion of turbulized dust-air mixtures," Combustion and Flame, vol. 123, no. 1-2, pp. 46–67, 2000.

18. U. Bielert and M. Sichel, "Numerische simulation von staubexplosionen in pneumatischen saug-flug-förderanlagen," VDI-Berichte, no. 1601, pp. 449–472, 2001.

19. K. Wörsdörfer, M. Sippel, J. Fuisting, and A. Kneer, "Möglichkeiten des Einsatzes numerischer Methoden im Explosionsschutz," VDI-Berichte, no. 1601, pp. 437–447, 2001.

20. V. P. Korobeinikov, I. V. Semenov, I. S. Menshov, R. Klemens, P. Wolanski, and P. Kosinski, "Modelling of flow and combustion behind shock waves propagating along dust layers in long ducts," Journal de Physique IV, vol. 12, no. 7, pp. Pr7/113–Pr7/119, 2002.

21. S. Zhong, A. Teodorczyk, X. Deng, and J. Dang, "Modeiing and simulation of coal dust explosions,"Journal de Physique IV, vol. 12, no. 7, pp. Pr7/141–Pr7/147, 2002.

22. P. Kosinski, R. Klemens, and P. Wolanski, "Potential of mathematical modelling in large-scale dust explosions," Journal de Physique IV, vol. 12, no. 7, pp. Pr7/125–Pr7/132, 2002.

23. A. Di Benedetto and P. Russo, "Thermo-kinetic modelling of dust explosions," Journal of Loss Prevention in the Process Industries, vol. 20, no. 4–6, pp. 303–309, 2007.

24. K. van Wingerden, B. J. Arntzen, and P. Kosi ski, "Modelling of dust explosions," VDI-Berichte, no. 1601, pp. 411–421, 2001.

25. B. J. Arntzen, H. C. Salvesen, H. F. Nordhaug, I. E. Storvik, and O. R. Hansen, "CFD-modelling of oil mist and dust explosion experiments," in Proceedings of the 4th International Seminar on Fire and Explosion Hazards, pp. 601–608, Londonderry, UK, September 2003.

26. O. R. Hansen, T. Skjold, and B. J. Arntzen, "DESC—a CFD tool

for dust explosions," in Proceedings of the 3rd International ESMG Symposium on Process Safety and Industrial Explosion Protection, Nürnberg, Germany, March 2004.

27. R. Siwek, K. Wingerden, O. R. van Hansen, et al., "Dust explosion venting and suppression of conventional spray dryers," in Proceedings of the 11th International Symposium Loss Prevention and Safety Promotion in the Process Industries, Praha, Czech Republic, May-June 2004.

28. T. Skjold, B. J. Arntzen, O. R. van Hansen, I. Storvik, and R. K. Eckhoff, "Simulation of dust explosions in complex geometries with experimental input from standardized tests," in Proceedings of the 5th International Symposium on Hazards, Prevention and Mitigation of Industrial Explosions (ISHPMIE ‹04), Krakow, Poland, October 2004.

29. T. Skjold, B. J. Arntzen, O. R. van Hansen, O. J. Taraldset, I. Storvik, and R. K. Eckhoff, "Simulating dust explosions with the first version of DESC," in Proceedings of the Symposium on Hazards XVIII: Process Safety—Shearing Best Practice, IChemE NW Branch Symposium, UMIST, Manchester, UK, November 2004.

30. T. Skjold, B. J. Arntzen, O. R. van Hansen, O. J. Taraldset, I. E. Storvik, and R. K. Eckhoff, "Simulating dust explosions with the first version of DESC," Process Safety and Environmental Protection, vol. 83, no. 2, pp. 151–160, 2005.

31. T. Skjold, B. J. Arntzen, O. R. van Hansen, I. E. Storvik, and R. K. Eckhoff, "Simulation of dust explosions in complex geometries with experimental input from standardized tests," Journal of Loss Prevention in the Process Industries, vol. 19, no. 2-3, pp. 210–217, 2006.

32. T. Skjold, "Review of the DESC project," in Proceedings of the 6th International Symposium on Hazards, Prevention, and Mitigation of Industrial Explosions (ISHPMIE ‹06), pp. 1–21, Halifax, Canada, August-September 2006, (Key-note paper).

33. T. Skjold, R. K. Eckhoff, B. J. Arntzen, et al., "Simplified modelling of explosion propagation by dust lifting in coal

mines," in Proceedings of the 5th Intenational Seminar on Fire and Explosion Hazards, The University of Edinburgh, Scotland, UK, April, 2007.

34. T. Skjold, "Review of the DESC project," Journal of Loss Prevention in the Process Industries, vol. 20, no. 4–6, pp. 291–302, 2007.

35. R. Weber, "Preisgekrönte Abhandlung über die Ursachen von Explosionen und Bränden in Mühlen, sowie über die Sicherheitsmassregein zur Verhütung derselben," Verhandlungen des Vereins zur Beförderung des Gewerbe-fleißes, pp. 83–103, 1878.

36. F. Hauert, A. Vogl, and S. Radandt, "Measurement of turbulence and dust concentration in silos and vessels," in Proceedings of the 6th International Colloquium on Dust Explosions, D. Xufan and P. Wolanski, Eds., pp. 71–80, Shenyang, China, August-September 1994.

37. P. Kosinski, R. Klemens, P. Wolanski, V. P. Korobeinikov, V. V. Markov, and I. S. Menʹshov, "Dust-air mixtures spreading in branched ducts," in Proceedings of the 18th International Colloquium Dynamics Exploration & Reaction System, Seattle, Wash, USA, 2001.

38. P. Kosinski and A. C. Hoffmann, "Modelling of dust lifting using the Lagrangian approach,"International Journal of Multiphase Flow, vol. 31, no. 10-11, pp. 1097–1115, 2005.
.

39. P. Kosinski, A. C. Hoffmann, and R. Klemens, "Dust lifting behind shock waves: comparison of two modelling techniques," Chemical Engineering Science, vol. 60, no. 19, pp. 5219–5230, 2005.

40. C. G. Ilea, P. Kosinski, and A. C. Hoffmann, "Three-dimensional simulation of a dust lifting process with varying parameters," International Journal of Multiphase Flow, vol. 34, no. 9, pp. 869–878, 2008. ·

41. K. Lebecki, J. Sliz, Z. Dyduch, and P. Wolanski, Critical Dust Layer Thickness for Combustion of Grain Dust, American

Institute of Aeronautics and Astronautics, New York, NY, USA, 1990.

42.   C. W. Kauffman, M. Sichel, and P. Wolanski, "Research on dust explosions at the University of Michigan," Powder Technology, vol. 71, no. 2, pp. 119–134, 1992.

43.   P. J. Austin, F. Girodroux, Y. C. Li, C. G. Alexander, C. W. Kauffman, and M. Sichel, "Recent progress in the study of dust combustion phenomena at the University of Michigan," in Proceedings of the 5th International Colloquium on Dust Explosions, pp. 211–214, Pultusk, Poland, April 1993.

44.   V. M. Boiko and S. V. Poplavski, "On the effect of particle concentration on acceleration of a dusty cloud behind a shock wave," in Proceedings of the 7th International Colloquium on Dust Explosions, GexCon AS, Bergen, Norway, June 1996.

45.   R. Klemens, P. Kosinski, and P. Oleszczak, "Mathematical modelling of dust layer dispersion by rarefaction waves," Archivum Combustionis, vol. 22, no. 1-2, pp. 3–12, 2002.

46.   A. V. Fedorov and Yu. A. Gosteev, "Quantitative description of lifting and ignition of organic fuel dusts in shock waves," Journal de Physique IV, vol. 12, no. 7, pp. Pr7/89–Pr7/95, 2002.

47.   A. V. Fedorov and N. N. Fedorova, "Numerical simulations of dust lifting under the action of shock wave propagating along the near-wall layer," Journal de Physique IV, vol. 12, no. 7, pp. Pr7/97–Pr7/104, 2002.

48.   F. Tamanini and E. A. Ural, "FMRC studies of parameters affecting the propagation of dust explosions,"Powder Technology, vol. 71, no. 2, pp. 135–151, 1992.

49.   J. A. H. de Jong, A. C. Hoffmann, and H. J. Finkers, "Properly determine powder flowability to maximize plant output," Chemical Engineering Progress, vol. 95, no. 4, pp. 25–34, 1999.

50.   N.O. Breum, "The rotating drum dustiness tester: variability in dustiness in relation to sample mass, testing time, and surface adhesion," Annals of Occupational Hygiene, vol. 43, no. 8,

pp. 557–566, 1999. ·

51. D. Dahmann and K. Möcklinghoff, "Das Staubungsverhalten quarzfeinstaubhaltige Produkte,"Gefahrstoffe- Reinhaltung der Luft, vol. 60, pp. 213–215, 2000.

52. T. Kletz, "Inherently safer design: avoidance better than control," in Proceedings of the 3rd World Seminar on the Explosion Phenomenon and on the Application of Explosion Protection Techniques in Practice, Flanders Expo, Gent, Belgium, February 1999.

53. R. K. Eckhoff, "Understanding dust explosions. The role of powder science and technology," Journal of Loss Prevention in the Process Industries, vol. 22, no. 1, pp. 105–116, 2009.

54. P. R. Amyotte and F. I. Khan, "An inherent safety framework for dust explosion prevention and mitigation," Journal de Physique IV, vol. 12, no. 7, pp. Pr7/189–Pr7/196, 2002.

55. P. R. Amyotte, M. J. Pegg, and F. I. Khan, "Application of inherent safety principles to dust explosion prevention and mitigation," Process Safety and Environmental Protection, vol. 87, no. 1, pp. 35–39, 2009. ·

56. C. Wilén, A. Rautalin, J. García-Torrent, and E. Conde-Lázaro, "Inerting biomass dust explosions under hyperbaric working conditions," Fuel, vol. 77, no. 9-10, pp. 1089–1092, 1998.

57. K. Schwenzfeuer, M. Glor, and A. Gitzi, "Relation between ignition energy and limiting oxygen concentrations for powders," in Proceedings of the 10th International Symposium Loss Prevention and Safety Promotion in the Process Industries, H. J. Pasman, O. Fredholm, and A. Jacobsson, Eds., pp. 909–916, Elsevier, Stocholm, Sweden, June 2001.

58. Dansk Fire Eater A/S, "INERGEN. Anlœgsbeskrivelse & Design," Report, Dansk Fire Eater A/S, Holte, Denmark, 1992.

59. M. Mittal, "Mathematical models for minimum explosible concentration of dusts," in Proceedings of the 5th International Colloquium on Dust Explosions, pp. 247–256, Pultusk, Poland, April 1993.

60. I. Hesby, Ignition of dust layers by metal particle sparks, M.Sc.

thesis, Department of Physics, University of Bergen, Bergen, Norway, 2000.

61.   J. Gummer and G. Lunn, "Ignitions of explosive dust clouds by smouldering and flaming agglomerates,"Journal of Loss Prevention in the Process Industries, vol. 16, no. 1, pp. 27–32, 2003. ·

62.   U. Krause and W. Hensel, "Zündgefahren lagernder Staubschüttungen—Neue Hilfsmittel für ihre Bewertung," VDI-Berichte, no. 1272, pp. 183–201, 1996.

63.   U. Krause and M. Schmidt, "Untersuchungen zur Zündung und Ausbreitung von Schwelbränden in Stäuben und Schüttgütern," VDI-Berichte, no. 1601, pp. 397–410, 2001.

64.   R. K. Eckhoff, "Dust explosion hazards in the ferro-alloys industry," in Proceedings of the 52nd Electric Furnace Conference, pp. 283–302, Iron and Steel Society, Nashville, Tenn, USA, November 1995.

65.   D. Lorenz and H. Schiebler, "Optische Temperaturmessung an Entladungsfunken im Hinblick auf deren Zündwirksamkeit bei Staubexplosionen," VDI-Berichte, no. 1601, pp. 653–667, 2001.

66.   E. Randeberg and R. K. Eckhoff, "Initiation of dust explosions by electric spark discharges triggered by the explosive dust cloud itself," in Proceedings of the 5th International Symposium on Hazards, Prevention and Mitigation of Industrial Explosions (ISHPMIE ‹04), Krakow, Poland, October 2004.

67.   E. Randeberg, W. Olsen, and R. K. Eckhoff, "A new method for generation of synchronised capacitive sparks of low energy," Journal of Electrostatics, vol. 64, no. 3-4, pp. 263–272, 2006. ·

68.   R. K. Eckhoff, W. Olsen, and O. Kleppa, "Influence of spark discharge duration on the minimum ignition energy of premixed propane/air," in Proceedings of the 7th International Symposium on Hazards, Prevention, and Mitigation of Industrial Explosions (ISHPMIE ‹08), vol. 1, pp. 44–53, St. Petersburg, Russia, July 2008.

69. E. Randeberg and R. K. Eckhoff, "Measurement of minimum ignition energies of dust clouds in the <1mJ region," Journal of Hazardous Materials, vol. 140, no. 1-2, pp. 237–244, 2007.·

70. R. K. Eckhoff and E. Randeberg, "Electrostatic spark ignition of sensitive dust clouds of MIE<1 mJ,"Journal of Loss Prevention in the Process Industries, vol. 20, no. 4–6, pp. 396–401, 2007. ·

71. H.-C. Wu, R.-C. Chang, and H.-C. Hsiao, "Research of minimum ignition energy for nano titanium powder and nano Iron powder," Journal of Loss Prevention in the Process Industries, vol. 22, no. 1, pp. 21–24, 2009.

72. G. Baudry, S. Bernard, and P. Gillard, "Influence of the oxide content on the ignition energies of aluminium powders," Journal of Loss Prevention in the Process Industries, vol. 20, no. 4–6, pp. 330–336, 2007.

73. M. Nifuku, S. Koyanaka, H. Ohya, et al., "Ignitability characteristics of aluminium and magnesium dusts that are generated during the shredding of post-consumer wastes," Journal of Loss Prevention in the Process Industries, vol. 20, no. 4–6, pp. 322–329, 2007.

74. L. Marmo and D. Cavallero, "Minimum ignition energy of nylon fibres," Journal of Loss Prevention in the Process Industries, vol. 21, no. 5, pp. 512–517, 2008.

75. Ø. Larsen, J. H. Hagen, K. van Wingerden, and R. K. Eckhoff, "Ignition of dust clouds by brush discharges in oxygen enriched atmospheres," Gefahrstoffe- Reinhaltung der Luft, vol. 61, no. 3, pp. 85–90, 2001.

76. R. K. Eckhoff, "A critical view on the treatment of combustible powders/dusts in the European 'Atex 100a' and 'Atex 118a' Directives," in Proceedings of the 3rd International ESMG Symposium on Process Safety and Industrial Explosion Protection, Nürnberg, Germany, March 2004.

77. R. K. Eckhoff, "Inadequate treatment of dust explosions and fires in 'ATEX'. A critical view on resulting standards for electrical apparatus," Bulk Solids & Powder Science &

Technology. In press.

78. Ch. Proust, "Laser ignition of dust clouds," Journal de Physique IV, vol. 12, no. 7, pp. Pr7/79–Pr7/88, 2002.

79. R. Klemens, P. Wolanski, and J. Klammer, "On unsteady flows of combustible dusty gases caused by a shock wave propagation," in Proceedings of the 8th International Colloquium on Dust Explosions, pp. 355–363, Safety Consulting Engineers, Schaumburg, Ill, USA, September 1998.

80. K. van Wingerden, G. H. Pedersen, G. H. Teigland, and R. K. Eckhoff, "Violence of dust explosions in integrated systems," in Proceedings of the 28th AIChE Annual Loss Prevention Symposium, American Institute of Chemical Engineers, Atlanta, Ga, USA, April 1995, Session no. 13 on Dust Explosions.

81. P. Holbrow, S. Andrews, and G. A. Lunn, "Dust explosions in interconnected vented vessels," Journal of Loss Prevention in the Process Industries, vol. 9, no. 1, pp. 91–103, 1996.

82. P. Holbrow, G. A. Lunn, and A. Tyldesley, "Dust explosion protection in linked vessels: guidance for containment and venting," Journal of Loss Prevention in the Process Industries, vol. 12, no. 3, pp. 227–234, 1999.

83. A. Vogl and S. Radandt, "Explosionsübertragung durch dünne Rohrleitungen," VDI-Berichte, no. 1601, pp. 575–594, 2001.

84. A. Vogl and S. Radandt, "Explosionsübertragung durch dünne Rohrleitungen," Tech. Rep. 05-9903, Forsch.gesellsch. angew. Systemsicherheit und Arbeitsmedizin, Mannheim, Germany, 2002.

85. M. Glor and K. Schwenzfeuer, "Einfluss der Sauerstoffkonzentration auf die Mindestzündenergie von Stäuben," in Dechema Jahrestagung, Wiesbaden, Germany, April 1999.

86. O. Devlikanov, D. K. Kuzmenko, and N. L. Poletaev, "Nitrogen dilution for explosion of nutrient yeast dust/air mixture," Fire Safety Journal, vol. 25, no. 4, p. 373, 1995.

87. E. Conde Lázaro and J. García Torrent, "Experimental research on explosibility at high initial pressures of combustible dusts," Journal of Loss Prevention in the Process Industries, vol. 13, no. 3–5, pp. 221–228, 2000.

88. R. K. Eckhoff, "Partial inerting-an additional degree of freedom in dust explosion protection," Journal of Loss Prevention in the Process Industries, vol. 17, no. 3, pp. 187–193, 2004.

89. F. Tamanini and J. V. Valiulis, "Improved guidelines for the sizing of vents in dust explosions," Journal of Loss Prevention in the Process Industries, vol. 9, no. 1, pp. 105–118, 1996.

90. E. A. Ural, "A simplified development of a unified dust explosion vent sizing formula," in Proceedings of the 35th Annual Loss Prevention Symposium, American Institute of Chemical Engineers, Houston, Tex, USA, April 2001.

91. CEN, "Dust explosion venting protective systems," European Union draft standard prEN 14491 (CEN/TC 305/WG 3/SG 5N, 27 February 2002) prepared by CEN/TC 305 'Potentially explosive atmospheres. Explosion prevention and protection', 2002.

92. R. Siwek, M. Glor, and T. Torreggiani, "Dust explosion venting at elevated initial pressure," inProceedings of the 7th International Symposium Loss Prevention and Safety Promotion in the Process Industries, pp. 57-1–57-15, SRP-Partners, Roma, Italy, May 1992.

93. T. Forcier and R. Zalosh, "External pressures generated by vented gas and dust explosions," Journal of Loss Prevention in the Process Industries, vol. 13, no. 3–5, pp. 411–417, 2000.

94. P. Holbrow, S. J. Hawksworth, and A. Tyldesley, "Thermal radiation from vented dust explosions,"Journal of Loss Prevention in the Process Industries, vol. 13, no. 6, pp. 467–476, 2000. ·

95. A. Harmanny, "Pressure effects from vented dust explosions," VDI-Berichte, no. 1601, pp. 539–550, 2001.

96. G. Li, X. Deng, W. Liu, et al., "Development of a quenching venting door (QVD)," in Proceedings of the 6th International

Colloquium on Dust Explosions, D. Xufan and P. Wolanski, Eds., pp. 530–534, Shenyang, China, August-September 1994.

97. A. Emde and B. Penno, "Einbindung der Sauerstoffverdrängung und des Kontraktionseffektes mit angepasstem Wiederstandsbeiwert Zeta bei der Entwicklung neuartiger Quenchvorrichtungen zur Explosionsdruckentlastung innerhalb von Räumen," VDI-Berichte, no. 1272, pp. 645–651, 1996.

98. E. A. Ural, "A simplified method for predicting the effect of ducts connected to explosion vents," Journal of Loss Prevention in the Process Industries, vol. 6, no. 1, pp. 3–10, 1993.

99. G. A. Lunn, "Institution of chemical engineers vent duct method applied to the VDI vent sizing technique," VDI-Berichte, no. 1601, pp. 513–526, 2001.

100. F. Tamanini and J. V. Valiulis, "A correlation for the impulse produced by vented explosions," Journal of Loss Prevention in the Process Industries, vol. 13, no. 3–5, pp. 277–289, 2000.

101. D. Crowhurst, "Explosion protection of industrial buildings," The European Summer School on Dust Explosion Hazards: Their Assessment and Control, organized by IBC Technical Services, in association with BMHB and IELG, Cambridge, UK, 1993.

102. K. Höppner, "Explosionsdruckentlastung von Gebäuden," VDI-Berichte, no. 1272, pp. 327–346, 1996.

103. F. Tamanini, "Dust explosion vent sizing. Current methods and future developments," Journal de Physique IV, vol. 12, no. 7, pp. Pr7/31–Pr7/44, 2002.

104. M. Silvestrini, B. Genova, and F. J. Leon Trujillo, "Correlations for flame speed and explosion overpressure of dust clouds inside industrial enclosures," Journal of Loss Prevention in the Process Industries, vol. 21, no. 4, pp. 374–392, 2008.

105. P. E. Moore, "Suppressants for the control of industrial explosions," Journal of Loss Prevention in the Process

Industries, vol. 9, no. 1, pp. 119–123, 1996.

106. K. Chatrathi and J. Going, "Effectiveness of dust explosion suppressants," in Proceedings of the 9th International Symposium on Loss Prevention and Safety Promotion Process Industry, pp. 1008–1017, Barcelona, Spain, May 1998.

107. A. Tyldesley, "Private letter to R. K. Eckhoff," November 1993.

108. P. E. Moore and R. Siwek, "Explosion suppression overview," in Proceedings of the 9th International Symposium Loss Prevention and Safety Promotion in the Process Industries, pp. 745–758, Barcelona, Spain, May 1998.

109. K. Chatrathi and J. Going, "Dust deflagration extinction," Process Safety Progress, vol. 19, no. 3, pp. 146–153, 2000.

110. K. Brehm, "Explosionsunterdrückung bei erhöhter temperatur," VDI-Berichte, no. 1272, pp. 261–272, 1996.

111. CEN, "Explosion suppression systems," European Union draft standard prEN 14373 (CEN/TC 305 WI 00305032, August 2001) prepared by CEN/TC 305 'Potentially explosive atmospheres. Explosion prevention and protection', 2001.

112. A. J. Morgan, The arresting of explosions to minimize environmental damage, Ph.D. thesis, Department of Mechanical Engineering, Brunel University, Uxbridge, UK, 2000.

113. A. Harmanny, "Duration of vented dust explosions," EuropEx Newsletter, vol. 23, pp. 5–9, 1993.

114. A. Harmanny, "Structural aspects related to explosion protection techniques," in Proceedings of the 2nd World Seminar on the Explosion Phenomenon and on the Application of Explosion Protection Techniques in Practice, EuropEx, Gent, Belgium, March 1996.

115. A. Harmanny, "Structural aspects related to explosion resistance of process buildings, structures and silos," in Proceedings of the 3rd World Seminar on the Explosion Phenomenon and on the Application of Explosion Protection Techniques in Practice, Flanders Expo, Gent, Belgium, February 1999.

116. G. Li, B.-Z. Chen, X.-F. Deng, and R. K. Eckhoff, "Explosion resistance of a square plate with a square hole," Journal de Physique IV, vol. 12, no. 7, pp. Pr7/121–Pr7/124, 2002.

117. K. Cybulski, Z. Dyduch, K. Lebecki, and J. Sliz, "Suppression of grain dust explosions with triggered barriers," in Proceedings of the 5th International Colloquium on Dust Explosions, pp. 437–447, Pultusk, Poland, April 1993.

118. U. Barth, "Explosionsgefahren managen—systematisch oder mit system?" VDI-Berichte, no. 1601, pp. 207–223, 2001.

119. U. Hesener, U. Barth, and B. Dyrba, "Erstellung von Explosionsschutzdokumenten anhand von Anlagenbeispielen der pharmazeutischen Industrie," VDI-Berichte, no. 1601, pp. 225–237, 2001.

120. M. M. van der Voort, A. J. J. Klein, M. de Maaijer, A. C. van den Berg, J. R. van Deursen, and N. H. A. Versloot, "A quantitative risk assessment tool for the external safety of industrial plants with a dust explosion hazard," Journal of Loss Prevention in the Process Industries, vol. 20, no. 4–6, pp. 375–386, 2007.

121. F. Alfert, "Cost comparison of dust explosion protection techniques available on the market," inProceedings of the 2nd World Seminar on the Explosion Phenomenon and on the Application of Explosion Protection Techniques in Practice, EuropEx, Gent, Belgium, March 1996.

122. H. Janssens, "Sicherheit zu einem erschwinglichen Preis!," VDI-Berichte, no. 1601, pp. 271–279, 2001.

Chapter 4

Road Tunnel Fire Safety and Risk: a Review

Jonatan Gehandler

SP Technical Research Institute of Sweden, Borås, 501 15, Sweden
Lund University, Lund, 221 00, Sweden

ABSTRACT

A review concerning road tunnel fire safety and risk is presented. In particular different perspectives and methods on safety and risk are discussed. Road tunnel fire safety usually involves high uncertainty and high-stakes decisions. Thus, a wider group of stakeholders and different types of knowledge should be included in the fire safety analysis and evaluation, than what is required by technical risk analyses. It is argued that the decision process should not be separated from the design and safety evaluation as they are strongly dependent and iterative processes. Decision theory can guide the design and decision process in negotiation with stakeholders. Key parameters for the decision can be analysed through a combination of functional requirements, societal and political values, safety engineering, safety factors and systems theory. By taking an organisational viewpoint, potential latent and active errors can be analysed and a good safety culture can be engineered. In order to improve the safety culture of truck companies, regulation ensuring proper maintenance, training and quality management may be necessary in a competitive global economy.

INTRODUCTION

Despite sometimes heavy regulation and sophisticated assessment methods, accidents continue to occur. A recent example is the Fukushima Daiichi nuclear power plant accident in 2011, which happened due to an earthquake followed by a 14 m tsunami wave. The plant had been designed for a 6 m wave despite that more severe waves had occurred in the past. Severe flooding have also happened near nuclear power plants before, why have we not learned (Epstein [2012]; Epstein et al.[2012])?

Several studies suggest that large uncertainties can be expected in a Quantitative Risk Analysis (QRA) (Amendola [1986]; Contini et al. [1991]; Lauridsen et al. [2001a],[b]; Fabbri and Contini [2009]), this is not least the case for road tunnels, where data is sparse and models for basic phenomenon such as fire behaviour, human behaviour and fire spread include rough assumptions, if they are at all considered (PIARC [2008]; Ferkl and Dix [2011]; Kirytopoulos and Kazaras K [2011]; Kazaras et al.[2012]; Rein et al. [2009]). Bjelland ([2013]) argues that the scientific framework within fire safety is too narrow. In order to improve fire safety, other methods and perspectives on safety and risk can contribute. This review article aims to explore different methods and perspectives concerning road tunnel fire safety and risk. A striking comment from a risk analysis assessor is that "what is actually quantified is the assessor's knowledge of the situation" (Contini et al. [1991]:146). This means that any model is limited by the assessor's understanding of road tunnels, traffic safety, human behaviour and tunnel fire dynamics, which will be the starting point of this review.

ROAD TUNNEL FIRE SAFETY

Setting the Scene

Fire requirements for tunnels and buildings in general are stated in the EU regulation on harmonised conditions for the marketing of construction products (CPR): *"The construction works must be designed and built in such a way that in the event of an outbreak of fire:*

- the load-bearing capacity of the construction can be assumed for a specific period of time;
- *the generation and spread of fire and smoke within the construction works are limited;*
- *the spread of fire to neighbouring construction works is limited;*
- *occupants can leave the construction works or be rescued by other means;*
- *the safety of rescue teams is taken into consideration."* *(CPR*[2011]*)*

Due to severe alpine tunnel fires in 1999 and 2001 the European Commission later released minimum requirements for road tunnel safety (EC [2007], [2004]) in support of the CPR. The EC-requirements cover administrative, organisational and technical aspects. Risk analysis as a method is highlighted for verification of safety. Due to the increased awareness of tunnel fire risk, several research projects where initiated including several tunnel fire tests (Ingason and Lönnermark [2012]; DARTS [2004]) and a study of the assessment of tunnel safety which further explored the use of risk analysis (Beard and Cope [2007]).

Tunnel Fire Dynamics

Despite that knowledge on tunnel fire dynamics now exist, enclosure fire dynamics is of importance, although some large differences exist (Ingason et al. [2015]). In enclosure fires, the heat and smoke is kept inside the enclosure and the availability of oxygen likely becomes a limiting factor. The size of openings will determine how large the fire can grow before it becomes ventilation controlled, i.e. controlled by oxygen supply (Karlsson and Quintiere [1999]). For enclosure fires, unburnt fuel can burn outside the enclosure openings as it is mixed with fresh air. When the fuel is surrounded by a gas mixture with less than approximately 13% oxygen, the fire will extinguish.

In tunnel fires fresh air is usually transported to the fuel along floor level which sustains the fire. Unlike enclosure fires all combustion takes place inside the tunnel and for ventilation controlled fires this can lead to nearly zero % oxygen further downstream. In tunnel fires the hot smoke initially rise and impinges on the ceiling, extends along the ceiling and gradually descends towards the floor as it is being cooled, see Figure 1. The amount of backlayering and the distance downstream that the smoke remains stratified is highly dependent on the ventilation conditions (Ingason [2012]; Ingason et al. [2015]).

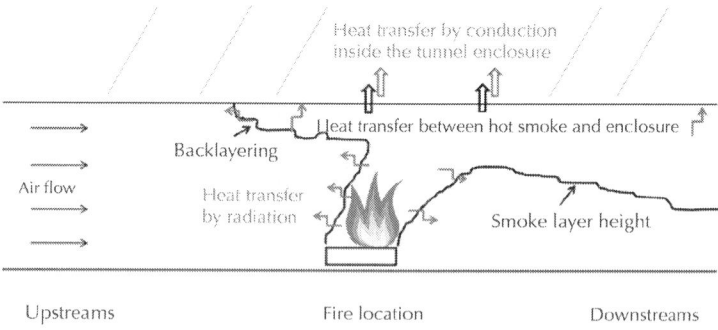

Figure 1: A schematic diagram over a tunnel fire introducing several important terms.

In recent years, a comprehensive theory on tunnel fire dynamics has started to develop. Fire parameters such as the temperature development, flame length, backlayering, visibility and gas concentrations can be calculated for tunnels with longitudinal air flow (Ingason et al. [2015]; Ingason[2012],[2008]). In tunnels with longitudinal or natural ventilation there is an air flow along the tunnel due to static and dynamic pressure differences. Transversal ventilation systems have air inlets and outlets along the tunnel length. With some minor modifications or limitations, a large part of the theory will also apply for tunnels with transversal ventilation. When a fire develops, buoyancy forces make the hot gases rise and spread along the ceiling depending on the tunnel inclination and initial ventilation. Depending on the air flow speed, the hot smoke and cold air mix and the smoke eventually becomes homogenously distributed in the cross-section downstream the fire. The first tunnel fire science study was performed by Thomas ([1958]) to study the effect of backlayering, when hot smoke travels upstream along the ceiling against the air flow, see Figure 1. Later Thomas ([1968]) introduced the concept of a critical air velocity needed to prevent backlayering. The critical air velocity will increase with the heat release rate (HRR) towards a constant value at around 3 m/s for most tunnels (Ingason [2008]; Ingason [2012]). The fire generates a resistance that increases with the fire size, called the throttling effect. Therefore, although 3 m/s will be sufficient to resist backlayering, the fan capacity has to be increased for increasing fire sizes (Vaitkevicius et al. [2014]).

The main fire load in tunnels concerns the vehicles that drive through it. A typical car has a fire growth rate corresponding to a fast[a] fire and a peak HRR at around 5 MW. A bus reaches around 30 MW and a heavy goods vehicle (HGV) between 20 and 200 MW with an ultra-fast[1] fire growth rate. For dangerous goods vehicles (DGV) there is no experimental data available although fires similar or worse than the HGV fire can be expected (Ingason and Lönnermark [2012]; DARTS[2004]). HGV and DGV fires can develop into catastrophic tunnel fires involving several vehicles with a ventilation controlled HRR between 300 and 700 MW (Ingason

[2003]). The heat release rate (HRR) of ventilation controlled fires will increase with larger cross-sectional area while fuel controlled fires (unlimited oxygen supply) will increase with decreased cross-sectional area due to increased heat transfer from the surrounding enclosure to the fuel (Ingason [2012]; Ingason et al. [2015]). Typically 2/3 of the HRR will be transferred by convection and 1/3 by radiation. If a sprinkler system is activated the convective part decrease to about 50%. The ventilation system is only affected by the convected HRR while the structure is exposed to both convective and radiative HRR Ingason and Li ([2014])).

ᵃFast, and Ultra-fast fire developments refers to the t-square model where the growth factor is defined as 0.047 and 0.19 (kW/s²) respectively (Karlsson and Quintiere [1999]).

According to Carvel et al. ([2001]) ventilation has a strong impact on the fire growth rate. In a more recent study Ingason and Li ([2010a], [b]) found the fire growth rate to increase linearly with the ventilation velocity. Also, depending on the type of fuel, ventilation conditions and fuel porosity the HRR can increase by a factor of 1–3 compared to free-burn tests if the ventilation is increased (Ingason [2005]; Lönnermark and Ingason [2007]). There is an upper limit to how much a material can burn per fuel area, therefore the HRR will reach a maximum level at which higher ventilation does not increase the HRR further (Ingason and Li [2010a], [b]). The ceiling height is an important parameter that receives limited attention when tunnels are designed, although the ceiling height together with the ventilation conditions is the most important parameters for the fire development. Another important factor is the geometry and the design of the vehicle on fire, e.g. containers or solid panels will significantly reduce the fire (Ingason et al. [2015]).

There are three mechanisms by which heat is transferred from one object to another: heat flux by radiation, conduction inside objects, and convection between hot air and objects, see Figure 1(Holman [2010]). The driving force of all heat transfer modes is the temperature difference. For the estimation of heat flux for tunnel applications, basic theory is presented in (Ingason et al. [2015]). Tunnel theory commonly ignores the effect of heat transfer through conduction, which means that the temperature development may seem independent of the thermal inertia of the tunnel structure. Decreased thermal inertia, e.g. insulation instead of concrete, will in general result in higher temperatures and faster fire growth rates; which has the potential to, for example, increase the risk of fire spread (Gehandler et al. [2014a], [b]; Gehandler and Wickström [2014]).

Modelling of fire in general as well as tunnel fires in particular is challenging as several basic mechanisms, e.g. combustion and fire spread, are poorly understood. Furthermore, modelling assumptions are numerous, e.g. the grid size, radiation model, turbulence model etc. In single comparisons between computer simulations and experimental data good results are often reported, e.g. (Hadjisophocleous and Jia [2009]). However, a round-robin study involving 11 independent teams reveals another picture (Rein et al. [2009]). A significant spread in the simulated results was found, despite the fact that each team received the same information of the fire test set-up that was to be modelled. The basic tunnel fire dynamics theory presented by Ingason ([2012]) seems to perform well in comparison with more advanced models (Nilsen and Log [2009]), and to offer sufficient precision for risk analysis (Gehandler et al. [2014a]).

Tunnel Accidents

In Europe, about 20 vehicle fires occur per billion vehicle km in tunnels (Nævestad and Meyer [2014]). Around 30% of all fires originate from HGV, despite that they only constitute 15% of the overall traffic volume (Ingason et al. [2005]; Nævestad and Meyer [2014]). An Austrian survey (Rattei et al. [2014]) covering the period 2006–2012 identified 38 car fires and 30 HGV and bus fires inside tunnels from national incident statistics, indicating HGV fires may have an even larger share of all fires. The rate of HGV and bus fires was 25 per billion km and the corresponding number for car fires was 4.2. A wide spread in the number of fires per vehicle km was found between different tunnels (Rattei et al. [2014]).

The most common causes for tunnel fires are collisions, overheating in combination with leakage or electrical failure, overheated bearings, brakes, tyres or engines. Collisions involving HGV or DGV are clearly overrepresented among the severe fires causing fatalities (Kim et al. [2010]). According to recent Norwegian statistics, injuries or fatalities result from traffic accidents rather than from fires (Nævestad and Meyer [2014]). In the Austrian survey only 7% of the fires were reported to have been caused by collisions, among which all were assumed to have included the entire vehicle. Out of 28 HGV fires caused by spontaneous ignition only 3 fires spread to the entire vehicle (Rattei et al.[2014]).

A Norwegian risk analysis estimated the expected loss of life from dangerous goods accidents to be less than 2% of the expected loss of life from normal traffic accidents (Lille and Andersen [1996]). An international survey of 1932 accidents during the transport of hazardous substances by road and rail found that the most frequent accidents were release of hazardous substances (78%) followed by fire (28%), explosion (14%) and gas clouds (6%). 63% of the accidents occurred on roads. Most accidents (75%) were caused by collision between vehicles. 3% or 13 accidents with hazardous materials took place in tunnels among which five were in road tunnels (Oggero et al. [2006]).

The major cause of collisions is driver error, according to a US study in 57% of cases and a UK study in 65% of cases. Adding all the cases when the road user was at least a contributing factor the numbers rise to around 95%. According to Oppenheim and Shinar ([2012]), traffic safety is more than the mere absence of accidents. We must go beyond accidents if we are to understand safe driving behaviour. Three error types can be distinguished as follows: #1 slips, i.e. right intention incorrectly executed, #2 lapses, i.e. failure to carry out any action at all, and #3 violations, i.e. deliberate deviation from accepted safe driving behaviour, e.g. speeding. Both slips and lapses relate to attention and memory failures. Lapses are of particular relevance to traffic safety as they relate to skill-based automatic behaviour. A mistake occurs when a driver intentionally performs an action that is wrong. Non-deliberative errors (lapses, slips and mistakes) may be reduced by training, memory aids, and better human-machine interfaces (Oppenheim and Shinar [2012]). Violations are best dealt with by trying to change users' attitudes by improving the overall safety culture.

Among environmental factors, high traffic density, narrow lane width, high horizontal curve grade, rising and falling gradients and limited lateral clearance are related to increased incident rates (Oppenheim and Shinar [2012]; Martens and Jenssen [2012]). In particular rising and falling gradients is highlighted to increase the number of HGV and bus fires in Austria (Rattei et al. [2014]). The area where the gradient goes from falling to rising is accident prone due to a likely abrupt change in speed (Martens and Jenssen [2012]), which is also confirmed by Norwegian tunnel incident statistics (Nævestad and Meyer [2014]). Tunnel entry portals have a high accident frequency, probably due to changing lighting conditions. The provision of traffic and safety information is necessary to improve driving behaviour and safety; but there is also a risk in providing too much information. In particular, information should be restricted 200 m before the tunnel entrance since then most drivers focus on the tunnel portal. It has been seen that many vehicles continue entering a tunnel even though traffic signals indicate the tunnel is closed, instead some kind of physical obstacle should be used (Martens and Jenssen [2012]).

Kim et al. ([2010]) analysed 69 tunnel fires and divided them in four incident categories:

- Single fires that do not spread to other vehicles. The majority (43) belong to this group. Only 11 caused fatalities.
- Single fires that propagate to neighbouring vehicles. All 5 fires in this category originated from HGVs and claimed fatalities.
- Collision fires limited to the vehicles that are involved in the collision. In 5 out of 7 cases fatalities occurred.
- Collision fires that spread to other vehicles which were not involved in the collision. 13 fires belong to this group and all claimed fatalities.

The analysis shows that fire spread is one of the key factors behind escalating consequences, both in terms of fatalities and tunnel downtime (Kim et al. [2010]). Ingason et al. ([2015]) offers a comprehensive theory on fire spread in tunnels. Fire spread is closely related to the HRR, gas temperatures, ceiling height, distance between neighbouring vehicles, flame length of the fire, and ventilation conditions. Tunnel fires can spread in a few minutes after the start of the fire (Kim et al.[2010]; Lönnermark [2007]). Fire spread in tunnels can occur through five main mechanisms (Ingason [2008]; Ingason et al. [2015]):

- Flame impingement due to flame tilt in the presence of a ceiling and due to the ventilation flow.
- Flame spread along the fire load.
- Spontaneous ignition of vehicles downstream due to increased temperature.
- Fuel transfer through leaking fuel tanks or debris downstream of the fire.
- Sudden deflagration.

Hansen and Ingason ([2011], [2012]) have developed a method for calculating the critical heat flux for ignition according to mechanism 3 above. Beard ([2006]) has developed a non-linear model called FIRE-SPRINT to identify the onset of instability with major fire spread according to either mechanism 1 or 3 above, see (Grant and Jagger [2012]; Charters [2012]) for an overview.

Despite the fact that fire spread and catastrophic fires involving multiple vehicles are key indicators of tunnel fire safety they are not accounted for among the most common QRA methods for tunnels (PIARC [2008]). Several parameters and systems can be used to reduce the risk of fire spread, such as reduced longitudinal ventilation speed, transversal ventilation systems, Fixed Fire Fighting Systems (FFFS) or manual extinction (Mawhinney [2011]; Ingason and Li [2010a]; Ingason [2012]). Transversal ventilation systems reduce the risk of fire spread outside the fire and smoke zone. In the near field of the fire, the risk of fire spread is similar to longitudinal ventilation. Transversal systems work effectively if enough fresh air is supplied from both sides of the fire (Ingason and Li [2010a]; Ingason et al. [2015]).

From a compilation of 20 fire fighting activities Kim et al. ([2010]) found that 15 were extinguished by the fire brigade or driver. It was further noted that fires caused by collision develop very rapidly and have a short time frame when fire fighters are able to approach them. Therefore FFFS is highlighted as a preventive measure to reduce catastrophic tunnel fires. From their study Kim et al. ([2010]) proposed the following four parameters for tunnel risk classification:

Allowance and amount of HGV and DGV

Bi-directional or uni-directional traffic flow

Tunnel congestion

Rescue service response time.

HGV and DGV represent a large risk as they constitute the fire load that potentially can lead to a catastrophic outcome. Bi-directional tunnels increase the risk of collision and make the evacuation process more difficult as road users are found both upstream and downstream of the fire. Highly congested tunnels have a higher frequency of collisions and greater risk of fire spread to nearby vehicles. Finally rescue service response time can significantly influence the ability for assisted egress and the potential for the fire service to approach the fire for manual firefighting.

Tunnel Fire Hazards

As for fires in general, exposure to smoke poses the main threat. During a large tunnel fire the hazards that an evacuee meet are numerous. Firstly, the visibility is impaired and the evacuee is forced to move through smoke. Within a few minutes, due to smoke irritants, the visibility is further impaired, furthermore, pain and breathing difficulties occur as irritants also affect the respiratory tract. After some further minutes of smoke exposure asphyxiant gases start to cause asphyxiation leading to confusion and loss of consciousness followed by death from hypoxia (Purser [2009]). Another hazard concerns the heat generated from the fire. Heat can be an issue for evacuation close to the fire, but foremost heat restricts rescue service intervention and can damage equipment or the tunnel structure, potentially leading to collapse or expensive renovation.

Most materials that burn are carbon-based. The end product of carbon-based fuels is mainly CO_2, H_2O and heat. CO_2 is a toxic asphyxiant gas in large doses. Depending on the fuel composition, temperature and ventilation conditions, other toxic products can be formed, e.g. CO or HCN (Blomqvist [2005]). Methods for quantifying fire hazards can be distinguished into limit-based and accumulative methods. In a limit-based method the gas concentration is estimated and compared with a limit value for each gas. If the limit is exceeded the evacuation has failed. By accumulative methods the accumulated effects from several asphyxiant gases are combined into a Fractional Effective Dose (FED) value. A FED value of 1.0 corresponds to the median of log-normal distribution of responses. A typical endpoint is incapacitation (Forster and Kohl [2012]; ISO [2012a]).

The risk of explosion most notably exists for transportation of gases that are liquefied by cooling or high pressure. An explosion occurs when the energy stored in the gas is released in a short time. In a full vessel almost all gas will be in liquid phase. A rupture in a full vessel leads to a sudden pressure drop to ambient causing the liquid to boil. The quick change from liquid to gas phase cause an increase in volume. Depending on the vessel temperature a blast wave can occur, if the evaporation is fast enough. This process is called, boiling liquid expanding vapour explosion (BLEVE). In an almost empty vessel much fuel will be in the gas phase. In this case a rupture causes an expansion of the pressurized vapour. The resulting blast depends on the temperature, type and amount of gas, and the dimensions of the tunnel. These two bursting vessel scenarios, without ignition, lead to high explosion loads in the zone around the bursting vessel (200–500 kPa), but is reduced after a couple of metres when the blast is directed along the tunnel axis, at around 100 kPa. Once the gas has expanded, ignition, e.g. by a spark or a hot surface, can occur if the gas-air mixture is within flammability limits. Depending on the speed of the flame front and expansion from combustion, a deflagration (10–800 kPa for HC-air mixtures) or detonation (1500–2000 kPa for HC-air mixtures) can take place. Of these, a deflagration in the order of 100 kPa is the most plausible scenario. Detonation is less likely as it requires instantaneous release of an almost empty tank. A pressure of 100 kPa or larger will cause direct casualties from the blast (Weerheijm [2014]).

Structural Behaviour

A number of past fires, for example the Channel tunnel fires and the Mont Blanc tunnel fire, show that fires pose a serious threat to the tunnel structure. There are four main types of tunnel constructions: cut and cover, immersed tube, drilled and blasted and bored tube tunnels. The dominating construction material is concrete. There are two main classes of concrete for tunnels: low-porosity (high-strength) and high-porosity concrete. For low-porosity concrete (often used in bored and blasted tunnels) the dominant failure process in tunnel fires is spalling, i.e. the explosive delamination of concrete. For high-porosity concrete in immersed and cut-and-cover tunnels the main failure mode is sagging of the roof due to loss of strength and expansion due to heat. Another threat for cut-and-cover and immersed tunnels is that the opposite, unexposed side, cracks. Measures to protect the tunnel integrity are, for low- and high-porosity concrete tunnels, either focused on withstanding fire exposure (fireproof concrete or insulation) or on fire suppression (Carvel [2005]; Carvel and Both [2012]).

The Eurocode offers general rules for structural fire design of concrete structures (CEN [2004]). It is generally sufficient to assume a fully developed ventilation controlled compartment fire with a uniform temperature distribution and to only verify individual members directly exposed to fire (Thomas [1986]). For this purpose, standardised testing of internal members using pre-defined time-temperature curves have been developed, e.g. the standard fire curve in EN 1363–1 and ISO 834, the hydrocarbon (HC) curve in EN 1363–2, or the Rijkswaterstaat (RWS) curve from the Dutch regulations. Members are classified according to the number of minutes that load-bearing capacity (R), integrity (E) or insulation (I) is ensured. A more performance-based alternative to the standardised fires is to develop a unique time-temperature curve given actual fire load and conditions, see for example the *Natural fire safety concept* (Sleich et al. [2002]). An attempt to develop a similar concept for performance-based tunnel design can be found in (Gehandler et al.[2014b]).

Human Behaviour in Fires

Key theories and concepts concerning human behaviour in fires were mainly developed during the 1970s and 1980s. More recently, interest in human behaviour during tunnel fires has started to develop, see (Shields [2012]; Noizet [2012]) for an overview. This research into human behaviour in tunnels has merely scratched the surface.

Social influence explains why we act differently to a fire threat alone and in groups, e.g. the apparent indifference of others can lead to passivity (Latané and Darley [1970]). The importance of social influence is believed to increase with decreasing distance to the nearest person and when the fire cue is unclear or uninformative (Nilsson and Johansson [2009]).

According to the *behaviour sequence model* the phases of evacuation are characterized by interpretation, preparation and action. The action in the last stage depends on previous stages. The activities people engage in to fulfil their role in any given situation are influenced by guiding principles or rules. When faced with a fire threat this *role-rule* attribute continues to guide the individual's behaviour (Canter et al. [1980]). The *affiliative model* suggests that people in a situation move toward familiar persons and places simply because they are familiar (Sime [1985]). In the *process model* the focus is shifted to human information processing and decision making. Earlier models for understanding human behaviour (such as the affiliative and role-rule model) can be used but in an iterative process. Two new concepts are introduced to describe the process. *Feedback in action* describes how people continuously act in response to new information rather than from an inert condition. *Effectance motivation* describes the continuous interaction of an individual with their environment to reduce uncertainties and ambiguities (Tong and Canter [1985]).

An important finding concerning human behaviour in fire is that people's reaction to an alarm is as important as the time it takes to physically move to an exit, if not more. In a lecture theatre evacuation study, two thirds of time from the onset of the alarm was spent not moving at all. Sime et al. ([1992]) therefore concluded that there is a disproportionate emphasis on time to move and exit flow rates in design standards and regulations.

The *theory of affordance* explains what affordances (perceived utility) an object such as an emergency door has on a person escaping. People perceive objects in terms of what they can offer or afford in relation to the fulfilment of their goal. Affordances can be divided into different categories depending on how they aid or support the user. Sensory affordance is the affordance of an object to be seen or sensed. Cognitive affordance supports understanding, such as how or why an object is used. Physical affordance supports the user physically, e.g. opening an emergency door. Functional affordance help users to achieve their goal (Nilsson [2009]).

As the understanding of human behaviour in fire in tunnels is limited, knowledge of human behaviour in buildings is of high value, although, differences between tunnels and buildings must be considered. The human-tunnel-vehicle system is different in many ways from that of human-building systems. Some of these differences are that road users are sitting inside a vehicle which in general is a familiar place and not on fire. Furthermore, the surrounding environment is an alien environment. The road user depends on visual impressions, since she cannot smell or hear much from the environment outside the vehicle. For buildings, user familiarity can sometimes be assumed, for tunnels user familiarity cannot be assumed. In particular, the notion of destination, person and property affiliation can explain why instructions to drivers often are disobeyed (Shields [2012]). Note that most studies on tunnel egress behaviour neglect differences in cognitive behaviour due to age and/or abilities (Noizet [2012]).

Emergency information is often provided for pedestrians. According to Shields emergency information should immediately be available for road users inside their vehicle. Especially considering that it has been noted in real tunnel fires that many road users stay in their vehicle (place of affiliation and familiarity) during an emergency. Emergency exits and signs should have sufficient affiliation to persuade the road user of the associated benefits. When driving through tunnels, signs, emergency doors or even the tunnel walls are hardly noticed, the side walls flash by due to the speed of the vehicle (Boer and van Zanten [2007]). The tunnel is seen in a flash and when tunnel users have to evacuate by foot they have no idea of the appearance of the tunnel.

In an evacuation experiment in the Benelux tunnel a truck fire was simulated to study human behaviour. In 6 out of 7 tests, motorists stayed in their cars until the first announcement. In one test motorists started to leave their cars immediately and others followed. In all seven tests the first announcement was sufficient to start the evacuation. One test showed extreme passivity by the motorists in the front who stayed in their cars even after they were engulfed with smoke. First after the second announcement did they react and commence evacuation. A common reason for not reacting to the incident was that no one else did anything. Another reason to stay in the car without reacting can be to stick to the role of being a motorist. As visibility decreases so do these social influences. This is believed to be part of the explanation as to why some motorists stayed in their cars being engulfed by smoke: they did not see the motorists leaving behind them (Boer and van Zanten[2007]).

Proulx and Sime ([1991]) investigated the efficiency of different communication systems for initiating evacuation in a Newcastle underground metro station. It was found that a regular alarm bell lead to a delayed evacuation or no evacuation at all. Although an alarm bell is supposed to mean 'evacuate the building', people seem to interpret the information as a system failure or a test. The will to reach the destination is so strong that everyone continued with their normal behaviour only slightly disturbed by the ringing of the bell. The response to evacuate was improved if staff members shouted at people to evacuate, or, even better, if a message was given on the public communication system. The fastest response was achieved when the message was timely and precise, e.g. a live voice describing what action is expected and why, and giving personal messages to people identified on the CCTV who had not started to evacuate. It is important that the message is clear, reliable, and easy to understand.

In a survey conducted on 151 firemen, truck drivers, regular drivers and student drivers, the management strategies in the event of a tunnel fire were investigated (Gandit et al. [2009]). The spontaneous response to a tunnel fire was to evacuate (40%), exchange information (35%), or to help others (13%), mainly through the use of a fire extinguisher. Of those who wanted to evacuate 50% looked for an emergency exit, 33% said they would move to the tunnel exit, and 17% towards the tunnel entrance. Gandit et al. ([2009]) concluded that although users are well aware of the safety devices, they do not use them automatically. Safety campaigns or a fire safety module in driver training courses could improve the situation to clarify why and how safety devices should be used (Gandit et al. [2009]).

There is a wide range of egress models available for buildings and an extensive review can be found in (Kuligowski et al. [2010]). As can be seen in the review above modelling of human behaviour is a challenging task as many parameters affect the complex decision-making process resulting in a wide range of behaviours. To account for this fact some models try to use artificial intelligence or probabilistic rules. Some models have been tested against fire drills or people movement experiments. One can expect a large operational uncertainty in applying these models, in particular with relation to tunnels.

Perspectives on Safety

Although this paper belongs in the technical science field it is also in accordance with Renn ([2008]), who believed that insights from other sciences, e.g. natural, psychology, economics, and cultural and social sciences, can enrich the understanding of safety and risk. The main paradigm for dealing with safety is risk analysis as developed from the technical science field, called technical risk analysis by Renn ([1998]). Similar to technical risk analysis, the economic concept of risk transforms physical harm and other effects into utilities. In contrast, a psychological perspective on risk reveals that we as individuals have a multidimensional concept of risk, which cannot be reduced to utilities, probabilities and consequences. A sociological perspective on risk tries to understand how the risk society works. A basic notion is that humans do not perceive the world with pristine eyes, but through perceptual lenses filtered by social and cultural meanings. Cultural theory seeks to make sense of the things humans do. Studying the origins of beliefs that guide risk-taking decisions reveals cultural patterns and different world views. This helps explain controversies concerning risk issues and explains why risk assessment cannot claim universal validity among all groups and cultures in society (Adams[2000]; Renn [1998]).

The scientific method can be defined in terms of the three characteristics: reductionism, repeatability, and refutation. The complexity of the real world is reduced in experiments whose results are validated by their repeatability and knowledge is built by refutation of hypotheses. The scientific method has been successful in many fields, however, complexity and social phenomena pose difficult problems. After having conducted case studies of fire safety engineering projects, Bjelland ([2013]) argues that the scientific framework for fire safety is too narrow to capture the essence of fire safety. In particular, reductionism leads to great simplifications in the treatment of complex systems and excludes critical issues that are difficult to quantify, e.g. human and organizational behaviour. This leads to an overemphasis of model concepts such as relative frequencies or causal structures. Bjelland ([2013]) highlights design science, systems safety and social constructivism as good compliments to the scientific method to broaden the view of relevant knowledge in the design process. In the design process, more emphasis should be placed on prior experience and tacit knowledge. Engineers should be allowed to creatively frame and reframe the problem in negotiation with stakeholders (Bjelland [2013]).

The method of systems is, unlike the scientific method, based on the idea that at certain levels of complexity there exist properties which are emergent at that level and which cannot be reduced to lower levels. An example of such a complex system is the human body with its organs, cells and DNA. At each level, e.g. that of organs, properties can be found that cannot be found at other levels (Checkland [1985]). Performance is controlled by the higher levels of system hierarchy. In order for this control to be effective there is the need for communication, feed-back and feed-forward about the state of the system (Bjelland [2013]).

Möller and Hansson find no less than 24 safety principles in the engineering literature, which are grouped in four categories as follows (Möller and Hansson [2008]):

- *Inherent safe design:* Potential hazards are excluded rather than just enclosed or coped with. In general this is the preferred solution if possible.

- *Fail-safe:* If the system does fail it should fail safely, or it should be fail-safe, i.e. internal components may fail without the system as a whole failing, or the system fails without causing harm. Defence in depth, reliability, and safety barriers are example of fail-safe concepts.

- *Safety reserves:* A system or construction is made strong enough to resist loads by a margin of safety to account for higher loads than foreseen, worse material properties than foreseen, imperfect theory of the failure mechanisms, possible unknown failure mechanism, and human error.

- *Procedural safeguards:* Procedures and control mechanisms are implemented to maintain safety. This includes safety standards, quality assurance, and training.

In general the efficiency of a safety measure decreases with increasing number above, i.e. inherent safety is more efficient than implementing procedures and safeguards. The Netherlands has adopted a policy for intrinsic infrastructure safety. To achieve decisions for intrinsic safety, a shared view of safety among all decision makers should emerge before safety objectives are evaluated against other objectives, e.g. economic (Rosmuller and Beroggi [2004]).

The safety circle in Figure 2 visualises different aspects of safety as a dynamic process of learning and improving. In any holistic safety approach all elements in the safety circle should be addressed, and it may be inefficient to only focus on one or a few. Pro-action is about eliminating the root causes, for example through training or design. Prevention is about reducing tunnel accident probabilities of crucial events, for example through reduced speed. Preparation concerns the management of emergencies. Mitigation (also called protection) is about mitigating the consequences of a tunnel accident. Intervention refers to the efforts of rescue teams. After-care actions are performed to quickly return to normal operation. Lastly, evaluation is about learning and constantly improving. Safety features that function early in the circle are in general most cost-effective (PIARC[2007]).

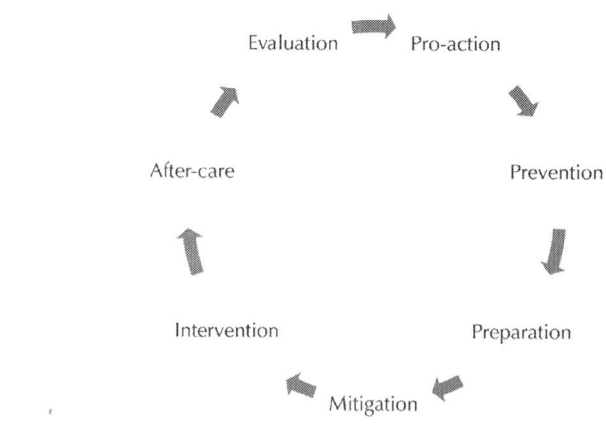

Figure 2: The safety circle (PIARC [2007]).

The five requirements stated in the CPR ([2011]) are largely consequence focused. One reason is that fire rules and regulation have developed as a reaction to occurred incidents. The fire is already assumed to have happened and regulations are designed to protect ourselves against future occurrences (IRCC [2010]). This is a reactive safety approach in contrast to a proactive approach. Consequently Malmtorp and Vedin ([2014]) find that about 80% of all safety measures aimed at tunnel safety focus on reducing consequences, despite the fact that preventive measures generally are more efficient. An overview of key terms and measures aimed at prevention and protection for tunnels is given by Beard and Scott ([2012]). Gehandler et al. ([2014a], [b]) argue that today's tunnel fire safety codes and standards do not cater to the complexity of modern multi entry and exit urban road tunnels. A suggested solution to account for both this complexity and the need for more proactive/preventive measures is to move to a performance-based design methodology and QRA (Gildersleeve and Sherlock [2014]; Malmtorp and Vedin [2014]).

However, both prescriptive and risk-based approaches have their positive and negative aspects. Prescriptive approaches contain a rich seam of knowledge and experience encapsulated in codes and guides, grounded in the real world, based on implicit risk but without explicit understanding of risk. A risk-based approach allows us to estimate the risk, although with several assumptions and considerable uncertainty, grounded more in models than in the function of the entire system in the real world (Beard [2004], [2012]). Fire models have the potential to be valuable and aid decision-making, but they also have limitations and can be used in ways which cause poor decisions to be made, see (Beard [1992], [1997], [2005]). Consequently Beard and Scott ([2012]) argue for a systemic approach where fire risk is seen as a product of the working of a system as a whole, and a healthy mixture of prescriptive requirements, qualitative risk assessment and quantitative risk assessment is applied (Beard [2012]).

Technical Risk Analysis

Due to the diversity of fields that deal with risk analysis, a wide spectrum of concepts and names are used. Sometimes the same words are used for different methods and sometimes different words are used for the same method; this is a fertile ground for confusion and misunderstandings (Kaplan[1997]). Central concepts for risk can be identified from the ISO ([2009a]) definition of risk which is: "effect of uncertainty on objectives" in which events, consequences and likelihood are key parameters. Uncertainty is the state of deficiency in information related to, understanding or knowledge of, an event, its consequences, or likelihood (ISO [2009a]). IEC/ISO ([2010]) defines the risk assessment process with the following phases: #1 *risk identification*, what can happen? #2 *risk analysis*, the consequences and likelihood of future occurrence is analysed, and #3 *risk evaluation*, decisions are made in relation to objectives and risks.

One way of classifying different models for risk analysis is by examining how uncertainty is treated. Six levels (from 0 to 5) are introduced by Paté-Cornell ([1996]). The success of analysis at various levels is dependent on resources, available knowledge, models and data. In some cases it does not make sense to perform an analysis at level 5 because there may not be any numerical models or data available. Uncertainty can also be treated in words by stating the gaps in knowledge, or through reducing the uncertainty in the system by making it more robust.

At level 0, the first step in risk analysis, risk identification is carried out. This can be sufficient for a strict zero-risk policy or for low cost decisions when the options are clear. Analysis at levels 1 and 2 consider a *worst* or *plausible worst* case and can be an option if this is sufficient to support a decision, e.g. to design for the maximum credible earthquake. The uncertainty in consequences is implicitly considered. This approach can be used in deterministic design procedures where scientific theories and empirical methods using conservative assumptions are used to evaluate the design as either successful or not (BS, [2001]). Analysis on Level 3 uses the best estimate or central value that reflects the most probable outcome and is often used in Cost and Benefit Analysis (CBA). An analysis on level 3 has a poor capability to capture the uncertainty of the outcome. (Paté-Cornell [1996]).

At levels 4 and 5, a Probabilistic Risk Assessment (PRA), or a QRA is performed. A distribution of probabilities is used in contrast to the previous deterministic approaches. This includes the worst case, plausible worst case, central values and a set or continuum of other cases. The output of level 4 is a risk curve over the likelihood for different consequences. This curve represents the uncertainty involved under the limitations of the method used and the assumptions made. At level 5 competing models and assumptions are taken into consideration and results in a distribution of risk curves providing an estimate of the inherent uncertainty of the risk measures (Paté-Cornell [1996]). PRA emerged from a reactor study on nuclear power plant safety (WASH-1400) in the mid-1970s (Stamatelatos et al. [2002a]). The introduction of the notion of 'scenario' contrasted with the deterministic practice current at the time, which was to only study reliability for a given design basis challenge, Level 1 or 2 above (commonly done in fire safety science where the term design fire is used to define the stress for the system in question). In contrast WASH-1400 studied several high consequence-low probability scenarios (Stamatelatos et al. [2002a]).

An informative definition of risk is the *set of triplets* definition (Kaplan and Garrick [1981]):

$$R=\{S_i,L_i,X_i\}_c, i=1,2,\ldots,N$$

The risk (*R*) is the comprehensive answer to the following three questions (Kaplan [1997]).

- What can go wrong? This gives all possible scenarios *Si*.
- What is the likelihood of each scenario *Si*? This gives *Li*.
- What are the consequences of each scenario *Si*? This gives *Xi*.

The brackets denote the set of triplets, i.e. the set of each scenario S_i with its likelihood L_i and consequence X_i and the subscript c implies that the set is complete, i.e. all relevant scenarios are evaluated. In practice the identified scenarios will never be complete as we do not know what we have not thought of (Beard [2002]); this calls for cautionary decision-making (Beard [2004]). No quantitative number or curve is "big" enough to capture the concept of risk. Scenarios and evidence also needs to be described in words since it is not possible to express everything in numbers (Kaplan[1997]). The aim to identify all scenarios including scenario descriptions, likelihood estimation and consequence estimation and description is according to Kaplan ([1991]) versatile and have worked well for several types of risk.

Another way to describe one or more risk scenarios is by logic diagrams, e.g. the bow-tie model (PIARC [2007]) or the crucial event model (Beard and Scott [2012]). The essence of these models is that causal factors come together to produce one or more events that then lead to consequences, see Figure 3. A causal factor can be of any nature, e.g. it may be a temporal event or condition such as 'fuel is present' or a latent condition. The causal factors can be further analysed in a Boolean fault tree with AND or OR gates representing the logic of how the causal factors produce the failure event. Likewise, the possible consequences from each event can be logically constructed in an event tree, see Figure 3. By applying probability theory to the fault and event trees the probabilities of the end states can be calculated (Stamatelatos et al. [2002a]).

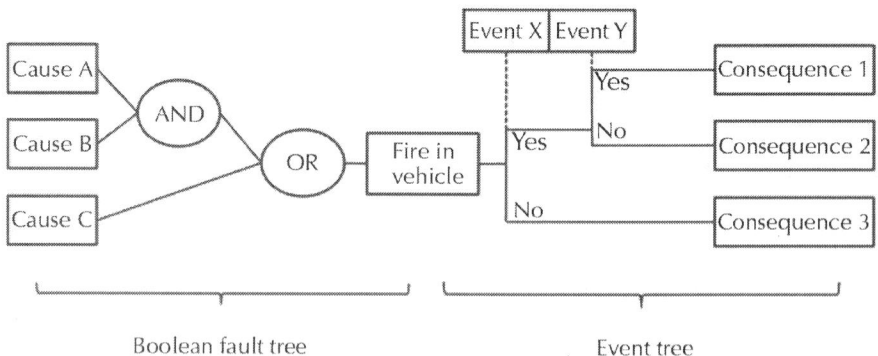

Figure 3: Fault tree and event tree.

An important qualitative result of fault trees is the minimal cut set (MCS) for top event failure to occur. An MCS is the smallest combination of basic events that result in the top event. Any MCS with only one basic event represents a single failure that alone can cause the top event to occur. These are often weak links in the safety chain. An MCS having events with identical characteristics are susceptible to common cause failures. Through a quantitative evaluation the dominant cut sets with the highest risk contribution can be identified (Stamatelatos et al. [2002b]). Although theoretically sound, it has proven difficult to model common cause failures (Renn [2008]). Nývlt et al. ([2011]) apply PRA on road tunnels using an unknown base probability. The logic is that, despite the fact that the probability of fire is unknown, it can be analysed and demonstrated to be well managed and mitigated. Similarly a logic tree approach is used by Beard ([1983]) where a reduction factor is calculated for various combinations of safety systems with unknown base probability.

Economic Perspective on Risk

Through an economic perspective on risk the physical damage is transformed into utilities. The objective yardstick for measuring utility is the amount of money someone is willing to pay for a change. By this transformation other aspects such as psychological or social effects can be measured besides physical harm. Furthermore risks and benefits can easily be compared as they are expressed in the common denominator of utility. Collective utility can be deduced by looking at past behaviour or through surveys. The economic perspective on risk conceptualizes risk as a cost factor that can be exchanged (Renn [1998]).

A controversial issue with the economic perspective on risk, e.g. CBA, is that *all* costs and benefits are translated into the single dimension of money, including e.g. life, which for many of us is considered to be incommensurable. Since resources are finite, society needs to make tough decisions when setting priorities between different life-saving alternatives (HSE [2001]). Mooney

([1977]) argues that valuation of human life for life safety decisions is an useful aid that complies with liberal democratic traditions and ensures rational decisions. According to social theory, human valuation and CBA are tools that are used by some groups in society to ease management, but lack validity among other groups (Adams [2000]).

Another controversial issue is how costs and benefits are to be compared over time. Economists have developed a widely accepted solution to this problem by discounting the future. According to Fischhoff and Kadvany ([2011]) it is questionable how well this applies to public decisions, e.g. future generations may not benefit from money that is saved today at the cost of the environment, and there is no obvious justification for discounting future lives.

According to Thomas ([1986]) the general objective of fire protection is to minimize the combined loss and cost of fire. An early application of utility theory on fire protective trade-offs was developed by Baldwin and Thomas ([1974]). In particular they were investigating the optimal combination between passive and active (sprinkler) fire protection. An important notion is that both active and passive protection may fail, there is in this sense no need to discriminate between the two modes of protection, and they both have a non-zero probability of failure. A balance has to be struck between the risk of failure, the ensuing damage, and the cost of reducing the risk or damage. A probabilistic approach for such evaluations is offered by Johansson ([2001]).

Risk Evaluation and Decision-making

Methodologies for risk evaluation and decision-making range from *hard* methodologies to *soft systems*methodologies. Hard methodologies are derived from the scientific method, characterized by reductionism, repeatability and refutation. At the other end of the spectrum are the soft systems methodology, e.g. by Checkland ([1985]). In a purely hard methodology, a considerable knowledge and understanding of the system is necessary. The method proceeds from problem to solution in a mechanical, orderly manner without any iteration. On complex and/or social systems the scientific method can be less successful, e.g. risk controversies where different actors have different values and objectives. The soft systems methodology is described as a never ending learning system that starts by expressing the situation where the perceived problem lies while not distorting the problem into a preconceived or standard form. Hard systems thinking (e.g. systems engineering and systems analysis) assumes that problems can be formulated as the making of a choice between alternative means achieving a known end (Checkland [1985]; Beard [2012]).

Beard and Cope ([2007], [2012]) proposed an intermediate methodology between the hard and the soft ends of the spectrum for tunnel fire safety. Such a methodology is the risk management process in IEC/ISO ([2010]), see Figure 4. Beard and Cope ([2007], [2012]) further presents a check-list concerning what a tunnel fire safety methodology should include, e.g. to make all assumptions clear, and to use an iterative process.

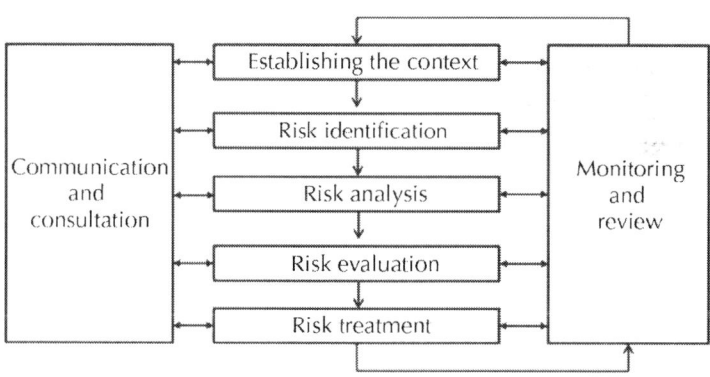

Figure 4: The IEC/ISO ([2010]) Risk management process.

Funtowicz and Ravetz ([1992]) argue that the limit of science is being reached for risk analysis involving ineradicable uncertainties in value-laden contexts. They argue that awareness of complexities in both the factual and the value-laden dimensions of the problems are necessary, which they call post-normal science. The gap between scientific expertise and a concerned public can be bridged by dialogue among all stakeholders. The democratization of the political life of modern societies means ordinary people can read, write, vote and debate. Funtowicz and Ravetz ([1992]) hope that a similar democratization of knowledge in society will take place, creating space for enhanced participation in decision making for common problems, which is necessary for meeting the challenges of modern times. This requires that the problem is framed in a way that acknowledges the different perspectives of the stakeholders, e.g. trustworthiness of managing institutions (Funtowicz and Ravetz [1992]; Wynne [1992]). Meacham ([2004a]) argues that fire safety design, involving modelling of fire and human behaviour with significant uncertainties, has reached the realm of post-normal science. This then requires the input from a broader group of concerned stakeholders in the decision process.

Risk analysis has achieved a more and more distinct and separate role in relation to decision-making and evaluation. This separation started in the 1980s when a National Research Council (NRC) report called the 'Red Book' proposed a division between analysis and evaluation arguing that this would remove overt policy values form the assessment part and ensure scientific expertise without value judgments (Vareman and Persson [2010]). An earlier NRC report had warned that it is difficult and sometimes unwise to separate analysis from evaluation. Some members of the committee felt that setting an ideal of value-neutral reporting of uncertainties is so unattainable that it distorts the analytical process (NRC [1982]). Fischhoff et al. ([1981]) argue that, although a distinction between facts and values enrich risk debates, such a distinction is often impossible to attain. The objectivity of a fact is always contingent on a correct statement of the problem. Beliefs concerning "facts" shape our values and those values in turn shape the facts we search for and their interpretation. According to Fischhoff et al., the search for an objective method such as risk analysis is doomed to fail and may obscure the value-laden assumptions that inevitably will be made (Fischhoff et al. [1981]).

Another consequence of the separation between risk analysis and decision-making and evaluation is that the search for the best decision is sometimes framed as an "acceptable risk" problem. In 1969 Chauncey Starr published a study aimed at finding a formula for determining whether risks were socially acceptable. He assumed that society revealed its preferences through the risks and benefits that are accepted from various hazards. The general rationale for acceptable risk is that if people accept one risk, they should accept all risks of the same order of magnitude measured in the same way. According to Fischhoff and Kadvany ([2011]) such comparisons are flawed in three fundamental ways:

by assuming that all risks can be defined by the same risk measure,

by assuming that risk decisions are about risk alone, and

to assume that accepted risks are acceptable.

According to Renn ([2008]), it is important to understand the central importance of benefits. Benefits are weighted versus risks and make them "acceptable". Risky decisions are not about risk alone. Rather they are a choice between options with different features, including the level of risk. When a technology is adopted, so is its entire package of features which means it is impossible to infer some level of acceptable risk. All relevant features must be included in risky decisions to find the right level of risk for each particular case (Fischhoff et al. [1984]; Slovic [2000]). As an example one may accept a large risk, such as smoking, if the benefit from smoking is perceived to be worth the risk; while one may reject a small risk from a chemical plant nearby that is perceived as bringing no benefits but noise and disturbance. In risk perception studies several more factors have been identified that affect how risks are perceived and judged by the public (Slovic [2000], [1987]). According to Otway ([1992]), attitudes towards technology as a whole, associated with risk, reveals a better understanding than the more narrow framing of risk perception.

It follows that one cannot and should not define risk in general terms suitable for all problems. Defining risk is a political act that expresses values regarding the relative importance of different possible adverse consequences for a particular decision (Fischhoff et al. [1984]). Whoever controls the definition of risk controls the rational solution (Slovic [2000]). Defining risk is a political and social act, determining what should be regarded as risk and how it is to be measured. The risk measure should be related to the decision context, e.g. if risk relates to an individual question concerning means of transportation the measure should reflect the whole journey from start to finish, and include the values of the concerned stakeholders (Holmgren and Thedéen [2003]).

Vrijling et al. ([1998]) argue that the degree of protection should be expressed in terms of acceptable risk. Additionally the choice of a certain technology and risk should be made in a cost-benefit framework. Since almost all studies on acceptable risk use two measures for acceptable risk Vrijling et al. use the same. One is the point-of-view of the individual who "decides to undertake an activity against direct and indirect benefits". The other measure considers if the benefits outweigh the risk for society. As the acceptable level of risk stands in relation to the benefits and voluntariness, the notion of acceptable risk needs to be flexible in relation to these aspects. Both the individual and societal level of risk needs to be "acceptable", i.e. below a defined threshold, or in relation to benefits and voluntariness. Society should represent the whole nation so that several local risks cannot add up on a national scale. Depending on the benefit and relation a person has with a given activity, a useful distinction in risk acceptability is often made between third party users, users/passengers, and employees. The framework rests on statistical accident data, similar to the study by Starr mentioned above, which shows that the individual risks can be ordered according to the generated benefit and voluntariness. To account for the different categories a policy factor, , is defined so that the individually acceptable probability of failure can be calculated accordingly (Vrijling et al. [1998]):

$$P_{fi} = \frac{10^{-4}\beta_i}{P_{d|f_i}}$$

Where $P_{d\backslash fi}$ denotes the probability of being killed in the event of an accident. In Table 1 policy factors ( ) for different types of activities are proposed, based on historical accident data.

Table 1: Different policy factors to account for different type of activities in terms of voluntariness and benefit (Vrijling et al.[1998])

Policy factorβ_i	Voluntariness	Benefit	Example and its individual risk	
100	Completely voluntary	Direct benefit	Mountaineering	10^{-3}
10	Voluntary	Direct benefit	Motor biking	-
1	Neutral	Direct benefit	Car driving	10^{-4}
0,1	Involuntary	Some benefit	Factory	10^{-6}
0,01	Involuntary	No benefit	LPG station	-

Gehandler

Gehandler *Fire Science Reviews* 2015 4:2, doi: 10.1186/s40038-015-0006-6

For social risk Vrijling et al. ([1995]) assume that individuals assess social risk on the basis of the events that occur within their circle of acquaintances. Assuming that each individual on average has 100 fairly close acquaintances, statistical data show that the recurrence of an accident claiming the life of one out of 100 acquaintances, is on the order of a human life span. They next use statistical accident data for different policy factors ( ) as above, which results in an activity that is permissible if it claims less than

$7*10^{-6}{}_i*$national population size

deaths per year. The model also includes a risk aversion index and a model to calculate locally acceptable risk from nationally acceptable risk.

For transportation risks it is noted that the applicability is questionable. One solution would require the definition of a standard unit length, but it is arbitrary what unit length is defined (Vrijling et al.[1995]). In a later article the framework is applied to road traffic where each vehicle is seen as an installation. As so many "installations" exist the current risk should not be acceptable according to this framework (Vrijling et al. [1998]).

Finally, the framework for acceptable risk proposed by Vrijling et al. ([1998]) aims at an economically optimal level of risk. The rationale is that the total cost for safer systems and expected total damage in monetary units is minimized. Vrijling et al. ([1998]) further underline that the three means they propose, i.e. individual and societal risk criteria and economical optimization, are just means to reach the goal of managed safety. The tools only measure some aspects of the entire system. The framework on risk acceptance proposed by Vrijling et al. is applied to tunnels by Arends et al. ([2005]). However, it is unclear how the method is applied in practice considering the lack of data which is also acknowledged by the authors.

The Tolerability of Risk (ToR) framework was developed by the UK Health and Safety Executive (HSE) in order to efficiently align decisions with policies and the preferences of UK citizens. Tolerability is a better word than acceptability since one does not really accept risks, although the practical implications are the same. In the HSE approach risks are characterized as unacceptable, tolerable or acceptable depending on the risk magnitude. In order for a risk to be tolerable it should be reduced to a level that is As Low As Reasonable Practicable (ALARP). CBA is the main tool to prove that a risk is ALARP (Bandle [2007]; Bouder et al. [2007]).

HSE apply the precautionary principle for hazards subject to high scientific uncertainty, which rules out lack of scientific certainty as a reason for not taking preventive action. A key point of the framework is to generate trust, therefore it is important to base the process on openness, transparency and stakeholder involvement (Guen [2007]; HSE [2001]). A drawback with ToR is that CBA and the ALARP principle do not consider how the benefits and risks are distributed, e.g. whether one person is benefitting grossly while many others are taking the risk (Fairman [2007]).

The approaches put forward by Vrijling et al. and HSE are the current dominating paradigm for risk evaluation and builds on utilitarian ideas where the collective is seen as a carrier of utility. The underlying rationale is that through a levelling of differences in cost per statistical life, financial resources can be used in a more cost-effective way allowing more lives to be saved (Hermansson[2005]; Hansson [2003]).

As already mentioned, this paradigm may not protect the individual from unfair risk exposure. Hermansson ([2005]) argues that risk management should acknowledge moral factors such as individual rights and fair risk taking. She also argues that the focus in risk management should shift from the outcome to the procedure for decision-making. Those affected by a risk decision should have the opportunity to be involved in a fair decision-making process. Public participation is a goal for democracy and a requirement for rational decision making (Renn [1998]; Hermansson [2010]).

Risk management implies value judgement on three levels: the choice of acceptability criteria, trade-off between criteria, and generation rational solutions. The dual nature of risk as a potential for physical damage and as a social construction demands a dual strategy for risk management. Public values and social concerns can identify the topics for risk management. Technical expertise can assess the magnitude and likelihood of risks, but public input is needed to set priorities and objectives (Renn [1998]).

Hermansson ([2007]) proposes a model that analyses ethical factors in risk issues. The model focuses on the ethical relationships amongst the three parties: the risk-exposed, the beneficiary, and the decision-maker. Seven questions have been developed to cover the ethical issues between the three risk parties (Hermansson [2007]):

- To what extent does the risk-exposed benefit from the risk exposure?
- Is the distribution of risks and benefits fair?
- Can the distribution of risks and benefits be made less unfair by redistribution or by compensation?
- To what extent is the risk exposure decided by those who run the risk?
- Do the risk-exposed have access to all relevant information about the risk?

- Are there risk-exposed persons who cannot be informed or included in the decision process?

Does the decision-maker benefit from other people's risk exposure?

In order to consider a wide range of concerns, Fischhoff and Kadvany ([2011]) put forward a British framework called 'concern assessment' that included a CBA and the six societal factors: familiarity, understanding, equity, dread, control, and trust. Each societal concern is measured with judgements allowing five levels for each attribute.

Bilson and Purchase ([2014]) employ a risk evaluation framework to tunnel safety that includes several ethical aspects. Utilitarian values are evaluated through a CBA. Duty ethics concern an evaluation of whether the required level of safety in terms of standards and regulation and societal expectations is achieved. Rights ethics concern an evaluation of the different perspectives of the owner, constructor and polititians (obviously the exposed should also be included here, see Hermansson above). Finally, virtue ethics is about finding a balanced decision that takes account of all relvevant factors.

In many fields such as nuclear safety, QRA has proven to be very successful to ensure and increase nuclear safety and aid cost efficient decision making (Apostolakis [2004]; Garrick et al. [2010]; Garrick [1998]). As pointed out by Apostolakis ([2004]), a QRA can improve safety decision making, but it is *not* a replacement for traditional safety methods or philosophies. QRA benefits include the logical and analytical consideration of thousands of scenarios, in-depth understanding of system failure modes, uncertainty quantification, identification of dominant scenarios so that resources can be wisely used (Apostolakis [2004]; Garrick et al. [2010]).

Uncertainty

Uncertainty is central to the concept of risk. Any decision involves uncertainty in several aspects, e.g. empirical parameters, decision variables, value parameters, model domain parameters or outcome criteria. Empirical quantities represent measurable properties of the real-world system being modelled, e.g. temperature or fuel price. Value parameters are quantities such as discount rate or value of life. Probability is a good way to express uncertainty, however, Morgan and Henrion ([1990]) argue that only empirical quantities should be represented by probability distributions. Uncertainty can also be treated by parametric sensitivity analysis, where the sensitivity in the output from deterministic changes to the uncertain quantity is examined, or by stating the knowledge base and made assumptions in words. Standard scientific practice deals with the technical level uncertainty. According to Funtowicz and Ravetz ([1990], [1992]) the methodological and epistemological levels of uncertainty should be dealt with qualitatively. The methodological level concerns systematic error and the epistemological level concerns ignorance.

Uncertainty in risk analysis is often classified into randomness (*aleatory*), representing variations in samples, or uncertainty due to inadequacies in the knowledge base (*epistemic*). When the evidence base is small the epistemic uncertainty is large. A third type of uncertainty is introduced by the risk assessor. Despite the use of the same models on well-defined problems a large operational uncertainty remains. Operational uncertainty includes the following factors, relevant for most risk analysis (Lauridsen et al. [2002]; Lauridsen et al. [2001a],[b]):

implicit or explicit assumptions about the nature of probability and choices among databases and within the same database,

system conceptualisation and hazard identification,

choice and use of models,

bias introduced by the context,

choice of boundaries, and

experience of the analysts.

Möller ([2006]) argues that any adequate concept of safety must include not only the measure of risk (including aleatory uncertainty), but also the measure of epistemic uncertainty. The epistemic uncertainty will be large for new or unknown risk since there are little or no statistical data. Then the risk should be judged to be high which is also how we intuitively perceive risks, e.g. we have an aversion against new or unfamiliar risks (Möller [2009]). If probability distributions were known, probabilistic models could be used to estimate the epistemic uncertainty. However, probabilistic distributions are seldom known to any accuracy. In particular it is difficult to correctly model the tails of probabilistic distributions. Unfortunately, in QRA and engineering design it is often the tails that matter. Svensson and Johannesson ([2013]) call design through the use of such uncertain relationships 'design by magic'. A more suitable method for estimating the epistemic uncertainty is Variation Mode and Effect Analysis (VMEA) which uses second order moment statistics which is more easily accessible. A more crude way to account for epistemic uncertainties is through the use of safety factors (Johannesson et al. [2013]; Svensson and Johannesson [2013]; Johansson et al. [2006]).

In general there are two issues to consider when using statistics in order to estimate the likelihood of an event. Firstly, the amount of statistics should be as large as possible, secondly they should be relevant to this particular site or system. These two objectives often work against each other. It is further important to consider that failure frequencies and accidents are not primarily caused by technical but organisational factors (Davidsson et al. [2003]). For tunnels, the collected data stretches over a few decades which mean the data relates to vehicles which have little relationship with modern vehicles in terms of heat release rates and other aspects of fire performance (Ferkl and Dix [2011]).

Risk Analysis Reliability

The subjectivity and inherent uncertainty of risk assessment can be considerable. Surprisingly few comparative experiments have been performed to give an idea of the accuracy of risk assessments which is very surprising as risk assessment is being widely used by scientists and engineers alike. According to the scientific method, any theory that does not yield comparable results when repeated by others on the same problem, should be refuted.

In the early eighties a systems reliability round-robin exercise was performed including several European teams on the auxiliary feed water system of a nuclear power plant. The exercise showed that modelling uncertainties were considerable and in some cases overwhelm data uncertainties due to different understanding of key concepts, e.g. common cause failures and human factors, and the analyst general judgements, e.g. use of data and information, interpretation of the system and use of different approaches/philosophies. This introduces a significant subjectivity in the assessment (Amendola [1986]).

In a round-robin exercise in 1990 eleven different teams of experts performed risk assessments on an ammonia storage facility given the same information and preconditions. The different methods applied, the different boundaries and hypothetical assumptions made for the accident sequences, and the different ways of calculating risk counters and presenting risk figures, made it very difficult to compare the final results on a common basis. Therefore, the authors argue that the comparative picture should not be taken as representative of the uncertainty in risk analysis in an absolute way. Large differences, one to several orders of magnitude, were found in the results and analysis by the different teams (Contini et al. [1991]).

The spread in results could be traced both to a large variability in event frequencies used, as well as consequence modelling. A large number of assumptions must be made to narrow down the infinite amount of scenarios to a manageable and understandable set that can be modelled. A multidisciplinary and collective procedure is recommended for the hazard identification phase to yield a more complete picture as this is a critical step in the analysis. Comparing the frequencies obtained from fault trees and statistics suggests that the technique of using fault trees to obtain failure frequencies is neither robust nor accurate. Even though the same model is used, the result could widely differ because the models were used differently. The authors conclude that transparency in terms of all the assumptions that are introduced in all steps of the risk analysis must be explained together with the result as they are strongly dependent (Contini et al. [1991]).

An interesting statement from one of the teams when operator reliability was assessed was that "what is actually quantified is the assessor's knowledge of the situation" (Contini et al. [1991]:146). The exercise did not allow for much interaction with any operators which partly explains the comment, however, engineering judgement is unavoidable as information is never complete and not all failure modes have been experienced by operators.

Another round-robin exercise on an ammonia storage facility was conducted in 1998–2001 by seven different teams. Again the intrinsic uncertainty present in risk assessment was significant and some of the main sources of uncertainty were identified as follows: the hazard identification phase, the estimation of scenario likelihood, and the calculation of consequences (Lauridsen et al. [2002]). These are three key aspects of the risk assessment process which were also identified in the earlier studies.

The uncertainties found were significant for decisions concerning land use. For example, the safe distance from a process industry differed in the worst scenario between 65 and 10000 m (Lauridsen et al. [2002]). Due to the intrinsic uncertainty in risk assessment, Fabbri and Contini ([2009]) argue that the resulting learning and increased understanding from performing QRA are more important than the actual risk estimate. This raise questions concerning todays tunnel fire design process since the risk analysis would often be carried out by an external consultancy and any lessons learned would not necessarily go into tunnel operational practice.

The reported uncertainties found in risk analysis may, however, be fundamental to any engineering model. In a round robin investigation covering 16 standard structural engineering calculations the results differed by several factors due to engineering modelling uncertainty (Fröderberg and Thelandersson [2014]). Due to several stochastic variables and limited knowledge, the modelling of fire and human behaviour for tunnels will be highly uncertain (Beard and Cope [2007]). Consequently any QRA on tunnel fire safety will be even more uncertain as large uncertainties concerning probabilities are multiplied with the consequence outcome of the modelling.

Beard ([1997], [2005], 2007) has offered recommendations for acceptable fire model use. In particular, the model itself needs to have the potential to be valuable. Further, a generally acceptable methodology of use which encourages the user to be explicit needs to be followed, and the user needs to be knowledgeable. Since the conditions for reliable and acceptable use of complex computer models for tunnel fires do not yet exist, several models may only be valuable in a qualitative sense rather than quantitative (Beard [2012]).

Fire Safety Engineering and Performance-based Design

Building on the ideas of risk analysis and risk-based design, Fire Safety Engineering (FSE) has evolved as a distinct research field in fire safety. One approach is to pursue the following steps. First, fire safety objectives are formulated qualitatively. Depending on the building and occupancy involved the fire safety objectives will be prioritized differently. The next step is to more precisely specify these goals according to the client's loss objectives. For example, one loss objective could be "no loss of life outside room of origin". For each objective one or more measurable functional requirement is formulated, and for each functional requirement, a performance criterion is specified. In other terms, the type and degree of fire stresses that equate to the stated loss objectives are specified. Such fire stresses could be a radiant heat flux or a rate of heat release. For example, the client loss objective of "no loss of life outside the room of origin", requires maintaining tenable conditions in all egress paths until all occupants outside the room of origin have been evacuated to safety. In quantifiable engineering terms tenability may be expressed as CO concentration, distance of the smoke layer above floor or visibility. Once the functional requirements and performance criteria are defined design proposals can be evaluated. The common method for doing this evaluation for buildings as well as tunnels is through a scenario analysis (deterministic risk analysis including one or a few scenarios) or a QRA (involving all identified and relevant scenarios). An acceptable design should fulfil the agreed loss objectives and performance criteria (ISO [2009c]; Meacham and Custer [1995]; ISO [2012b]; PIARC [2007]). Gehandler et al. ([2013], [2014b]) have developed a performance-based design guide for road tunnel fire safety.

In scenario analysis a number of characteristic scenarios are selected to test the trial designs. The selection of scenarios is critical. The potential number of scenarios is infinite and a manageable set has to be identified. Each fire safety design objective has its own set of challenging scenarios. It is important that the resulting design solution is conservative (ISO [2006]). The consequences for each scenario are evaluated against a pre-defined criterion. The scenario-based risk analysis is also a suitable method for the planning of tunnel emergency response measures (PIARC [2008]).

The basis for deciding a performance-based acceptable level of risk is that the available safe escape time (ASET) is larger than the required safe escape time (RSET) by a margin of safety. The objective is often that all occupants should be able to escape without experiencing or developing serious health effects. The margin of safety depends on the chosen fire scenarios, the uncertainties in the calculations, and the fire safety objectives (ISO [2009b]).

Bjelland and Njå ([2012]) find that current practice of ASET/RSET analyses in the Norwegian building industry are done to confirm that chosen solutions are sufficient while the analyses themselves have limited constructive value for engineering design. Out of 75 examined projects, none contained evaluations of more than one design alternative.

According to Babrauskas et al. ([2010]), the ASET/RSET concept is flawed precisely because it is used, as the example above illustrate, to verify fire safety to an "acceptable level", rather than to maximise fire safety. Roughly half of all deaths and 2/3 of the injuries could be prevented if more time was available for escape. To try to save these people another method or concept seems to be necessary. Consequently Babrauskas et al. ([2010]) advise against the idea to define quantitative criteria as a measure of acceptable safety. Instead they propose a safety factor approach to be used.

FSE advocate performance based design in favour of prescriptive regulations. Standardisation aims to standardise the design and resist unique solutions. Both approaches have cost-efficiency and safety as an argument for their rationale. FSE proponents argue that if the solution is tailored to the situation the construction will be more effective and cheaper while standardisation argue that if solutions are standardised the wheel does not have to be reinvented (Johnson [2012]; Ruijter[2012]). Ruijter recognise that standardisation is not possible for all aspects of a tunnel but highlight safety demands and operational processes as highly appropriate for standardisation. One practical advantage would be that all tunnels and safety equipment would look and work the same way. Preferably the safety equipment should look and work the same in all regions within which drivers operate, e.g. in all of Europe.

Human Error and Organisational Accidents

One drawback with technical risk analysis is that organizational aspects are excluded (Renn [1998],[2008]). The starting point of this section is that, in order to improve safety, human error has to be understood. Three types of human strategies in problem-solving can be distinguished: skill-based, rule-based, and knowledge-based (Reason [1997]). If possible the fast and skill-based strategy will be applied. If no suitable skill-based strategy is found, the problem is compared to similar rules and if a suitable rule which has been used several times before with success is identified it is applied. If no rule-based strategy can be found that works, an analytical and knowledge-based solution is generated. Depending on which problem-solving strategy that is used different error-types can be identified: slips and lapses are connected to the skill-based strategy, and mistakes to the other two strategies (Akselsson [2011]). From a cognitive perspective, due to the mechanism of the human mind and its response to the environment, errors are unavoidable and should be seen as a consequence rather than a cause (Reason [1997]).

Since evacuation in tunnels is an unfamiliar activity, a skill-based problem-solving strategy will not be adopted. A rule-based strategy may be adopted through finding the similar event of evacuating a public or private building during exercise or real emergencies. However, it is likely that no past situation and successful strategy is matched with the current situation which means that a more time-consuming knowledge-based strategy is initiated. The error type concerned with rule-based and skill-based strategies is mistakes, i.e. wrong action such as staying in the vehicle is carried out. A driver inside a tunnel needs all possible help to speed-up the knowledge-based strategy so that the correct action to evacuate is performed as fast as possible. Since we are aware of these factors surrounding evacuation, it is a design error not to support the road user correctly. As Reason says, human error is a consequence, not a cause. Due to the difficulty in achieving a fast human response in the event of fire, it may be wise to give obligatory information or even training in driver licence courses.

Reason ([1997]) further distinguishes error by *active error* whose effects are felt immediately and *latent error* whose adverse consequences may lie dormant within the system for a long time. In general active errors are associated with front-line operators while latent errors are caused by decision makers and management separated in time and space. Detailed analyses of accidents in complex systems such as nuclear power plants or industrial sites reveal that latent errors pose the greatest threat to safety. Examples of latent failures relevant to fire safety are the corroding sprinklers of Piper Alpha and the inability to realise the fire risk in London metro (Reason [1990]; Akselsson [2011]).

Reason ([1997], 1990) offers a theoretical framework for accidents in complex systems. According to Reason, production systems (e.g. mass transportation) share several basic elements in common and can be generalised into the following five components:

- decision makers (e.g. designers and high-level managers),
- line management (e.g. maintenance, training),
- preconditions (e.g. reliable equipment, safety culture),

- productive activities (i.e. integration of human and mechanical elements), and
- defences (i.e. safeguards)

There is a flow from #1 to #5: decisions from decision makers (#1) are implemented by line managers (#2), this in turn affect the preconditions (#3) and later the actual performance in delivering the right product at right time (#4). The defences (#5) prevent foreseeable injury, damage or outages in the product activities. Feedback loops return feedback to the line management and decision makers. Operators carry out their duties such as maintenance and production activities managed by the line management and affected by the preconditions of the workplace.

All of the five mentioned components of production can have human contributions to failures. These failures can either become latent system failures or they are active failures. For the corresponding component number above, errors can be categorized accordingly:

- fallible decisions (latent),
- line management deficiencies (latent),
- psychological precursors of unsafe acts (latent),
- unsafe acts (active), and
- Inadequate defences (latent & active).

According to Reason ([1990]) system's accidents have their primary origin in fallible decisions made by designers and high-level management. The key factors that contribute to fallible decisions are safety and production goals which in turn are affected by money, equipment, personnel, and available time. An accident occurs when an unsafe act is committed in the presence of a potential hazard for which latent failures from decision makers, psychological precursors, and the defence coincide. Reason uses the word unpredictable to describe the coincidence of latent and active errors that cause an accident, which suggest quantification is not very meaningful. Similarly a review of 1000 shipping accidents concluded that accidents resulted from highly complex coincidences which could rarely be foreseen by the people involved (Reason [1990]).

Lately there have been several incidents in Norwegian road tunnels, e.g. Gudvanga 5 August (UPI[2013]) and Storsand 22 August (Adressa [2013]), one reason is that poorly maintained foreign trucks increase the risk of fire. In the incident in Gudvanga a fire started in a Polish truck. One political and societal factor responsible for this is the strive for larger markets and globalization. Transportation markets are enlarged within EU which means low salary countries enter the market of richer countries. The tough competition decreases the resources for safety, training and maintenance. This is a typical example of the struggle between production and protection in organizations (Reason[1997]). It may be time for stricter regulation aiming at proper maintenance, quality management systems, and defensive driving culture. Another issue is that foreign drivers may not understand the culture, language and road signs in the country where they drive.

The human layer can be seen as the last layer of protection. Since we know the human element is dynamic and will always change, latent failures that are allowed in the other layers of protection will eventually be exposed by the human layer and cause an accident. Since we knew that the human element was variable, it was really the latent failures which caused the accident. To increase safety, latent failures must be minimized, identified, and monitored so that barriers can be constructed before them (Reason [1997]).

In the tunnel-vehicle-driver system the front line operators are the drivers themselves. Reason highlights, among others, the importance of the front-line operators. From the review on road safety by Oppenheim and Shinar ([2012]) it is obvious that the human factor is a key factor causing road accidents. In particular this has to do with lapses, i.e. failure to respond to a threat. For tunnel fires the human factor is also a contributing factor. Fires may start as a result of crashes, but they can also start while driving e.g. overheated brakes or engines. Technical failures can be due to poor maintenance, poor design, or bad luck. Note that there is a human element behind the poor maintenance and poor design as well. A Canadian truck company managed to reduce the number of incidents through creating a culture of safety within the fleet.

A Trucking philosophy was established which was to serve as a reminder of the drivers responsibilities. It was displayed around the facilities and on material distributed to the drivers. A safety committee was established where safety was discussed. Training on defensive driving was given bi-annually to the drivers. A rating system was introduced with personal incident ratings for each driver. The number of incidents was almost halved in four years (Menzies [2007]). This shows that a safer culture can be engineered and improve road and tunnel safety.

This indicates that we can only reach high road tunnel safety by reaching out to all citizens, to establish a national and even international culture of road safety. As is noted by Holm ([2007]), the Swedish society has a poor road safety culture. It is difficult for authorities to take decisions such as lowered speed or traffic barriers aiming at improving safety when the citizens living there work against them and mainly prioritize high availability and accessibility. Several campaigns have been performed to alter the public perception into a more safety oriented perspective, not least safety belt, keeping speed limits, and 'drinking equals no driving' campaigns. Cultural beliefs and habits are naturally transferred from older to younger generations, as it becomes part of how we do things, resulting in safer roads in the long run.

Taking an even larger perspective, a certain company may have international organisations and national governments, regulators and associations on higher level, whose decisions affect their activities. Many nested levels of decision-making are thus involved in how, for example, a hazardous process is dealt with. Unfortunately this is seldom studied as a whole, instead several research disciplines study different levels so that, for example, management theories are independent of the context of a given organization. But the study of decision-making cannot be separated from the study of the social context and value system in which it takes place (Rasmussen and Svedung [2000]).

To account for the nested levels of decision-making, Rasmussen and Svedung ([2000]) propose a framework called *proactive risk management*. The first step towards proactive risk management is to ensure operation within the design envelope. The mechanism generating the actual behaviour of decision-makers at all levels has to be understood. Their values and objectives as well as their need for information and feedback have to be clarified. This involves a top-down communication of values and objectives and a bottom-up communication of actual state of affairs. The method and framework necessary to maintain a high level of safety is a Total Quality Management (TQM) system (Rasmussen and Svedung [2000]).

No matter how many improvements that are suggested from different efforts, improvements are dependent on the organisation's ability to learn and to improve in reality. To support the process of learning a TQM system and Deming's cycle, which aims at constant improvement through an iterative cycle: plan, do, study, and act (PDSA) can be applied (Akselsson [2011]).

Tunnel operators have many tasks. They are monitoring the traffic flow and traffic situation, detect disturbances, closing the tunnel if necessary, communicate with users, communicate and assist the emergency service, reporting and evaluation. Since incidents and especially larger fires are rare, training and exercises of such situations is very important. Another parameter that affects their performance is their cognitive load depending on business and the complexity of their tasks. Cognitive over-load and under-load is believed to worsen performance (Martens and Jenssen [2012]).

Systems Safety

A central concept for understanding risk is that of a *system*, which Beard ([2012]) defines as: *any entity, conceptual or physical, which consists of interdependent parts*. In contrast to a purely reductionist approach, risk concerns the system as a whole, as it functions in reality. Since systems change and tunnel risk is complex and multi-faceted, any analysis will be incomplete (Beard and Scott [2012]). This is, according to Hollnagel ([2010]), captured through the terms tractable and intractable systems. Typically a tractable system is simple to describe with few details, principles of functioning are known, the system does not change while being described and it is independent of other systems. An intractable system is the opposite. A metaphor for a tractable system is a clockwork and a metaphor for an intractable system is teamwork. According to Hollnagel ([2010]) most socio-technical systems are intractable. Current approaches to safety assume the system to be tractable and furthermore make the following assumptions (Hollnagel [2011]):

Systems are well designed and scrupulously maintained

The procedures that are provided are complete and correct People behave as they are expected to, and more important, as they are trained to System designers have been able to foresee and anticipate every contingency

Under those assumptions humans are clearly a liability and a threat. Example of frequently used methods to control this liability includes training, standardisation, rules and regulation. This approach represents an ideal but is not practically achievable. The two main reasons for this are that most systems are intractable and that performance variability is inevitable (Hollnagel [2011]). As an example, Lutz ([1993]) examined 209 safety-related software errors concerning two space crafts. He found that the main root causes for errors were discrepancies between documented requirement specifications and actual requirements needed for correct functioning of the system, and misunderstanding of the system interface with the rest of the system.

Acknowledging that precise procedures and instructions are not attainable, an alternative approach for intractable systems considers adaptation to meet functional goals as a necessary process. In this way performance variability is seen as an asset rather than a threat. In fact, according to Hollnagel, performance variability is on the whole the reason why socio-technical systems works as well as they do. Assessment methods must be able to capture the duality that human performance both can enhance and detract safety. From such a viewpoint systems work because (Hollnagel [2011]):

people can learn to identify and overcome design flaws and functional glitches, people can recognise the actual demands and adapt their performance accordingly,

when procedures must be applied people can interpret and apply them to match the conditions, and finally people can detect and correct when something goes wrong or when it is about to go wrong, and hence intervene.

This is a more realistic description of work as actually done, rather than imagined, hence systems that are real rather than ideal. Since both failure and success depends on performance variability, failure is seen as opportunities for learning (Hollnagel [2011]).

Kirytopoulos and Kazaras ([2011]) argue that QRA of tunnels suffer from the following limitations.

The probability of a fire starting in a tunnel cannot be reliably calculated.

The complexity of tunnel accidents is too large.

Large difficulties and assumptions in assessing human behaviour.

The influence of management and organizational aspects are often neglected despite that they are believed to be the key factor for safety in socio-technical systems.

Therefore, they propose a systems theory approach and a method called STAMP. In STAMP the accident model is viewed as interconnected networks rather than sequential events as in QRA. Furthermore, much analysis is made on management and organization to make it function well. It is largely a proactive approach to assess whether the organization is effective enough to keep the system within safety constraints. STAMP will not result in the same output as QRA why they could be used in parallel. (Kazaras et al. [2012]).

The STAMP assessment process for tunnel safety proposed by Kazaras et al. ([2012]) begins by identifying hazardous system states and translate them into safety constraints. To achieve the safety constraints, a safety control structure over components and paths of control and feedback loops is defined (socio-technical). By using the safety control structure inadequate control actions are identified and used to determine necessary safety functions (Kazaras et al. [2012]).

Santos-Reyes and Beard ([2012]) take a systemic approach to tunnel fire safety management. In their framework the systemic approach is compatible with QRA. The tunnel fire safety management model is also used by the authors as a template for comparison with an actual real world system in order to improve the existing management system (Santos-Reyes and Beard [2011]; Santos-Reyes and Beard [2006]; Santos-Reyes and Beard [2003]).

Design

Following the traditions of natural and technical sciences, safety engineering becomes an activity of structuring goals and performance criteria into mathematical language (Bjelland [2013]; Meacham and Custer [1995]). This approach assumes well-structured problems and leads to a narrow view on what is considered as relevant knowledge. In contrast, design science can be seen as a reflective conversation with the situation that highlight the skills and experience that designers and engineers bring to situations of uncertainty and value conflicts. Important designer skills are creativity, the ability to frame the design problems in different ways and to structure different solutions based on previous experience. Design processes are not linear and the stakeholders' goals and values will be conceptualized and refined during the design process (Bjelland [2013]).

According to Hollnagel ([2006]) there has been a technological bias in design in the sense that design for technology came first and design for humans at a distant second. However, putting the human at the centre of things is just as inadequate as machine-centred design since one part of the system is seen as opposed to the other. Design should therefore embrace a function-centred view and be problem-driven. For tunnels this means we should study the joint tunnel-vehicle-human system, and design should further the purposes or goals of this joint system, i.e. to be in control in a dynamic environment (Hollnagel [2006]).

Ruland et al. ([2012]) takes a function-cantered systems approach to road tunnels. They incorporate Systems Engineering (SE) and other safety tools into the whole design process. SE highlights both validation (are we building the right thing according to the road users need?) and verification (are we building it right, are all specifications correctly implemented?). In the Netherlands the infrastructure authority use SE as a working method to administrate their contracts. Their design process includes the following steps. Clear and accurate specification of what the system is, does, and should handle. The specification process starts from top requirements and specifies lower level system requirements into finer and finer detail. Each subsystem, and its effect on the system as a whole, is analysed from four perspectives: Reliability, Availability, Maintainability and Safety (RAMS). Scenario-driven tools such as scenario analysis or table-top exercises validate that the system and subsystems offer the required functionality. Eventually, the specification is detailed enough to start the realisation. Each specification step is then validated and verified as the design is being realised, from smaller components to larger system parts (Gehandler et al. [2012]; Ruland et al. [2012]; Ruland and Snel[2010]).

Safety Culture

Pidgeon ([1997]) views culture as a system of symbols or meanings through which a given group understands the world. "Such a culture is itself created and recreated as members repeatedly behave and communicate in ways which seem to them to be natural, obvious and unquestionable, and as such will serve to construct a particular version of risk, danger and safety." (Pidgeon [1997]:7). A good safety culture can be supported by the following factors (Choudhry et al. [2007]; Pidgeon[1997]):

A shared care and concern for hazards.

Realistic and flexible norms and rules about hazards.

Continual reflection upon practice.

Work with attitudes and behaviour.

Management commitment (allocation of resources, to "walk the talk", inspections).

Employee involvement (empowerment, involvement of employees).

Promotional strategies (mission statements, slogans)

Training and seminars.

Special campaigns (e.g. health week).

According to Reason ([1997]) a safe culture can be engineered through organizational characteristics like structures and systems which becomes a collective practice that an organization can have. The safety culture is the engine that drives the organization towards better safety, and its power is derived from 'never forgetting to be afraid'. It consists of an informed culture, i.e. right information is collected and spread. It is dependent upon a reporting culture which relies upon a just culture, i.e. treating each other in a way that is morally right. Finally a safety culture must draw the right conclusions and have the will to implement changes, i.e. it must also be a learning culture.

Decision Theory

According to Fischhoff and Kadvany "the foundations of risk lie in decision theory, which articulates concepts whose emergence must have begun with the first human thought about uncertain choices" (Fischhoff and Kadvany [2011]:2). The logic of decision-making is to choose the option that promises most of what you want. Meacham ([2004a]) has written a review on decision-making for fire risk problems. Most decision theories are based on Bernoulli's concept that choice depends on the likelihood of various outcomes and on the utility of those outcomes to the decision-maker, e.g. Expected Utility Theory. Social Choice Theory is a concept of rationality for synthesizing preferences among individuals affected by the decision, e.g. consensus building that takes into account primarily the facts and values of those participating in the development of fire safety regulation. Once a regulation is in place CBA plays a more central role when fire risk decisions are required for specific projects, see (Johansson [2001]). In this case the decision-maker is less concerned with the "social good" than providing an "acceptable" level of safety at a minimum cost (Meacham [2004a]). For tunnels the situation is quite different. Usually it is the state that develops fire safety regulations and plans, builds and owns tunnels. In all cases, safety and the social good have a high priority.

Decision making is fundamental to all fields. A general model called PrOACT which is applicable to any decision is offered by Hammond et al. ([1999]). The method consists of eight elements: problem, objectives, alternatives, consequences, trade-offs, uncertainty, risk tolerance, and linked decision. The essence of the method is to divide and conquer. By systematically breaking down the problem into smaller parts focus can be directed to the most critical aspects. In order to focus on the most important parts the process should rather be cyclic, i.e. iterative, than sequential.

The way the problem is stated frames the decision and determines what can be regarded as solutions, in fact posing the right problem drives everything else. By questioning the problem statement the root trigger can be identified and constraints that narrow the range of considered alternatives can be removed. Objectives specify the goal of the decision, and give the direction to strive for. Objectives can be identified by specifying all the concerns that the decision must address. To reach the fundamental objectives, ends are separated from means (potential decision alternatives) through why-questions. What-questions clarify each objective and increase the understanding of how to reach it (Hammond et al. [1999]).

Alternatives are the different courses of action available to choose from. The decision can be no better than the best alternative. The objectives can identify decision alternatives by asking "how?" for each objective. Next, consequences from each alternative are evaluated for each objective. Often objectives conflict with one another, which is why trade-offs are inevitable. If an alternative is dominated by another on practically all objectives it can be eliminated. For tougher trade-offs the even swap method can be used to eliminate objectives for which all alternatives are equally good. In this sense both alternatives and objectives can be eliminated iteratively, resulting in more manageable decisions (Hammond et al. [1999]).

The future is always uncertain and different outcomes will be more or less certain. Hammond et al. ([1999]) propose the usage of risk profiles to capture information about uncertainty. A risk profile answers the following questions.

Which are the key uncertainties?

What are the possible outcomes of these uncertainties?

What are the chances of occurrence of each possible outcome?

What are the consequences of each outcome?

This risk profile share many similarities with a risk analysis. The last two questions are included in the triple definition commonly used in risk analysis (Kaplan [1991]; Kaplan and Garrick [1981]). If the safety analysis is complemented by an explicit analysis of epistemic uncertainties, all four questions above are covered. Möller ([2009], 2008) argue that epistemic uncertainty should be included in any concept of safety. The practical experience of Hammond et al. ([1999]) is that that all decisions involve uncertainties, but most uncertainties do not influence consequences enough to matter. By identifying the few uncertainties that influence the decision, analysis can be performed where it matters. The resulting risk profile can be expressed in the form of a decision tree where each fork represents an uncertainty and the branches the outcomes and likelihoods.

Depending on the risk tolerance, risk profiles will seem more or less beneficial. This can be quantified by a desirability scoring, which in essence has close similarities with expected utility used in economics. The desirability curve will reveal whether we in this case are risk aversive, risk neutral, or risk seeking. There are several means to re-shape the risk profile into a more desirable one, e.g. through risk sharing, to seek risk-reducing information, to diversify the risk, to hedge the risk, or to insure against the risk (Hammond et al. [1999]).

Current FSE and risk analysis practice does not seem to acknowledge the decision problem context and the overall aim to find the best decision. Hammond et al. ([1999]) offers a comprehensive list of error types in decision making in which examples of poor fire safety decision-making from this review are given in parenthesis:

working on the wrong problem (e.g. rather than discussing safety, it is often a discussion of whether the design is better or worse than a prescriptive solution. This may include great emphasis on finding an appropriate prescriptive reference building (Bjelland [2013]).),

failure to identify key objectives (e.g. to aim for an 'acceptable' design rather than 'saving as many lives as possible' (Babrauskas et al. [2010]).),

failure to develop good and creative decision alternatives (e.g. to only develop and evaluate *one*design alternative (Bjelland and Njå [2012]).),

overlooking crucial consequences (e.g. to ignore the risk of fire spread and multiple vehicle fires.),

giving inadequate thought to trade-offs (e.g. by having rigid regulations that does not allow for trade-offs),

disregarding uncertainty (e.g. to disregard considerable uncertainty in the Rogfast road tunnel risk assessment (Bjelland and Aven [2013]).),

failure to account for relevant risk tolerance, and failure to plan ahead when decisions are linked over time.

It is argued that these errors can be reduced if the decision-problem is acknowledged and systematically dealt with.

DISCUSSION

The different methods and perspectives of this review highlight different aspects of safety and risk. They all have the potential to be valuable for road tunnel fire safety. No single method or perspective can claim universal validity. Only through combining several methods and perspectives can an efficient approach to managing road tunnel fire safety be achieved.

Tunnel fire safety is largely a low probability-high consequence risk issue. Small fires (5–20 MW) are seldom any issue for life safety or business continuity. Larger fires occur rarely, but can mean both loss of lives as well as long tunnel closure and expensive repair costs. The uncertainty in estimating probabilities and modelling of fire and consequences is considerable. Decision stakes are often high in terms of investment costs and the risk of longer tunnel closure and life safety. The methodological framework of the fire safety community is too narrow for these problems to be efficiently addressed (Bjelland [2013]). The limits of post-normal science are being reached (Meacham [2004a]) and a broader group of stakeholders should now be included in the decision process. The realm of relevant knowledge should be extended to include other sciences, concepts and methods of ensuring safety. A risk decision is not merely about risk or cost, although these are two important factors, each risk decision have challenges, uncertainties and factors that society value, this should be reflected in the decision process and guide the process and trade-offs.

In practice this could mean putting more trust in tacit and prior experience rather than formalized risk assessment, although, for example, fire modelling should follow standard procedures of good practice. The concept of safety factors for different functional parameters should be used in qualitative and quantitative ways instead of an overall quantitative risk measure. Depending on data and modelling uncertainty, quantitative methods may only be valuable in a qualitative sense. In the words of Svensson and Johannesson ([2013]) this is a move towards enlightened engineering rather than design by magic. In the creative and cyclic process the design group frames and reframes the problem and potential solutions in negotiations with stakeholders. The design process should further be function centred and problem driven. Performance-based design offers a good starting point with a complete set of basic goals, objectives and functions of the tunnel system, e.g. (Gehandler et al.[2014b]), but authorities and engineers need to make the best out of the new freedom offered by performance-based design. Several examples in this review show that better solutions or safer design does not come for free. Good examples can be found in the Netherlands with intrinsic safety efforts in early decision making and Systems Engineering in the design process, which enforces verification and validation of needed functions.

It is argued that decision-making should not be separated from design and evaluation as they are strongly dependent and iterative processes. Decision-making is fundamental to most reviewed methods, therefore we should acknowledge that we are dealing with a decision problem. Then the tools for decision-making, see section (Decision theory), can be used to structure the problem, to remove constraints and biases, to identify the basic objectives and potential solutions, to evaluate solutions and to perform trade-offs. It is likely a few factors will show up as the most important ones to evaluate further. Then a set of suitable methods that evaluate these aspects can be selected, taking into account their limitations, uncertainty and strengths. In light of new understanding along the iterative process the problem and potential solutions are reframed. Guidelines exist that have the potential to improve fire safety decision-making, e.g. Beard ([2012]) and Meacham ([2004a], [b]).

Most tunnel fire safety measures focus on protection despite the fact that pro-active and preventive measures in general are more efficient. However, all aspects of the safety circle, see Figure 2, need to be included for safety to be managed efficiently. Intrinsic safety and fail-safe design are two efficient engineering principles. Acknowledging the nature of human error and the importance of a well-functioning organisation, latent errors should be analysed, reduced and controlled. A good safety culture within the tunnel organisation should be engineered. A TQM system can ensure improved safety during operation in the long run. Systems thinking can further remove safety constraints and faulty design in the real socio-technical system.

CONCLUSIONS

Road tunnel fire safety concerns high uncertainty and high-stake decisions. This means the decision process should include a wider group of stakeholders and include different types of knowledge, e.g. prior experience, safety engineering, decision theory, systems theory, social science and design science.

It is argued that the decision process should not be separated from the design and safety evaluation. Instead decision theory should be used to structure and drive the process; to identify the basic objectives, alternative solutions and key uncertainties, and prioritize resources for analysis where they matter the most.

An efficient pro-active safety measure would be to improve the safety culture of professional drivers and truck companies. Regulation ensuring proper maintenance, training and quality management may be necessary in a global competitive economy.

ACKNOWLEDGMENTS

This study was funded as part of SP:s centre of excellence on tunnels and underground facilities (SP Tunnel). The author would like to thank his colleges Haukur Ingason, Anders Lönnermark, Margaret McNamee, Francine Amon and supervisor Håkan Frantzich for valuable comments and support. The author would also like to express his gratitude to the reviewers of this paper, their comments have enhanced this paper's quality.

REFERENCES

1.   Adams J (2000) Risk. Routledge, London.
2.   Adressa. Omkom i ulykke i Storsandtunnelen - Et vogntog tok fyr inne i tunnelen på E39. 2013. adressa.no: http://www. adressa.no/nyheter/sortrondelag/article8144363.ece   (In Norwegian). Accessed 15 Jan 2014

3.   Akselsson R (2011) Människa, teknik, organisation och hantering av risker (In Swedish). Institutionen för Designvetenskaper, LTH, Lund, Sweden.

4.   Amendola A (1986) Uncertainties in systems reliability modelling: Insight gained through European Benchmark exercises. Nucl Eng Des 93(2–3):215-225 http://dx.doi.org/10.1016/0029-5493(86)90221-9

5.   Apostolakis GE (2004) How useful is Quantitative risk assessment? Risk Anal 24(3):515-520

6.   Arends BJ, Jonkman SN, Vrijling JK, van Gelder (2005) Evaluation of tunnel safety: towards an economic safety optimum. Reliab Eng Syst Saf 90(2–3):217-228 http://dx.doi.org/10.1016/j.ress.2005.01.007

7.   Babrauskas V, Fleming JM, Don Russell B (2010) RSET/ASET, a flawed concept for fire safety assessment. Fire Mater 34(7):341-355 http://dx.doi.org/10.1002/fam.102

8.   Baldwin R, Thomas PH (1974) Passive and active fire protection — The optimum combination. Fire Technol 10(2):140-146 http://dx.doi.org/10.1007/bf02642517

9.   Bandle T (2007) Tolerability of Risk: The Regulator's Story. In: Bouder F, Slavin D, Löfstedt RE (eds) The Tolerability of Risk: A New Framework for Risk Management, Earthscan, London. pp 93-104

10. Beard AN (1983) A logic-tree approach to the St Crispin Hospital fire. Fire Technol 19(2):90-102 http://dx.doi.org/10.1007/bf02378564

11. Beard A (1992) Limitations of computer models. Fire Saf J 18(4):375-391 http://dx.doi.org/10.1016/0379-7112(92)90025-8

12. Beard AN (1997) Fire models and design. Fire Saf J 28(2):117-138 http://dx.doi.org/10.1016/S0379-7112(96)00082-3

13. Beard AN (2002) We don't know what we don't know. In: 7th International symposium on fire safety science, Worcester, MA, USA. pp 765–775

14. Beard AN (2004) Risk assessment assumptions. Civ Eng Environ Syst 21:19-31 http://dx.doi.org/10.1080/102866003 10001605489

15. Beard AN (2005) Requirements for acceptable model use. Fire Saf J 40(5):477-484 http://dx.doi.org/10.1016/j.firesaf.2004.10.003

16. Beard AN (2006) A theoretical model of major fire spread in a tunnel. Fire Technol 42:303-328

17. Beard A (2012) Decision-making and risk assessment. In: Carvel R, Beard A (eds) Handbook of Tunnel Fire Safety, 2nd edn. ICE Publishing, London. pp 635-648

18. Beard A, Cope D (2007) Assessment of the Safety of Tunnels - Study. Science and Technology Options Assessment. European Parliament, Brussels.

19. Beard A, Scott P (2012) Prevention and protection: overview. In: Carvel R, Beard A (eds) Handbook of Tunnel Fire Safety, 2nd edn. ICE Publishing, London. pp 67-88

20. Bilson M, Purchase A (2014) Determining benefits of fixed fire fighting systems in road tunnels - A risk-based approach. In: Ingason H, Lönnermark A (eds) Proceedings from the Sixth International Symposium on Tunnel Safety and Security (ISTSS 2014), Marseille, France. SP Technical Research Institute of Sweden, pp 475–484

21. Bjelland H (2013) Engineering Safety with Applications to Fire Safety Design of Buildings and Road Tunnels. University of Stavanger, Norway, Stavanger.

22. Bjelland H, Aven T (2013) Treatment of uncertainty in risk assessments in the Rogfast road tunnel project. Saf Sci 55(0):34-44 http://dx.doi.org/10.1016/j.ssci.2012.12.012

23. Bjelland H, Njå O (2012) Interpretation of safety margin in ASET/RSET assessments in the Norwegian building industry. Paper presented at the PSAM11 & ESREL 2012, Helsinki,

24. Blomqvist P (2005) Emissions from Fires - Consequences for Human Safety and the Environment. Report 1030. Lund University, Lund, Sweden.

25. Boer LC, van Zanten DW (2007) Behaviour on tunnel fire. Springer Berlin Heidelberg, Berlin, Heidelberg.

26. (2007) The Tolerability of Risk: A New Framework for Risk Management. Risk Society and Policy series. Earthscan, London.

27. (2001) Application of fire safety engineering principles to the design of buildings-Code of practice. British standards Institution, London.

28. Canter D, Breaux J, Sime J (1980) Domestic, Multiple Occupancy, and Hospital Fires. In: Canter D (ed) Fire and Human Behaviour, John Whiley & Sons, Ltd. pp 117-136

29. Carvel R (2005) Fire protection in concrete tunnels. In: Carvel RO, Beard AN (eds) The Handbook of Tunnel Fire Safety, Thomas Telford Publishing, London. pp 110-126

30. Carvel R, Both K (2012) Passive fire protection in concrete tunnels. In: Beard A, Carvel R (eds) The Handbook of Tunnel Fire Safety, 2nd edn. ICE Publishing, London. pp 109-126

31. Carvel RO, Beard AN, Jowitt PW, Drysdale DD (2001) Variation of heat release rate with forced longitudinal ventilation for vehicle fires in tunnels. Fire Saf J 36(6):569-596 http://dx.doi.org/10.1016/S0379-7112(01)00010-8

32. (2004) Eurocode 2: Design of concrete structures - Part 1–2: General rules - Structural fire design. European committee for standardization, Brussels.

33. Charters D (2012) Control volume modelling of tunnel fires. In: Beard A, Carvel R (eds) Handbook of Tunnel Fire Safety, 2nd edn. ICE Publishing, London. pp 347-364

34. Checkland P (1985) Systems Thinking. Systems Practice, Wiley, Chichester.

35. Choudhry RM, Fang D, Mohamed S (2007) The nature of safety culture: A survey of the state-of-the-art. Saf Sci 45(10):993-1012 http://dx.doi.org/10.1016/j.ssci.2006.09.003

36. Contini S, Amendola A, Ziomas I (1991) Benchmark Exercise on Major Hazard Analysis. Discussion of the Results and

Conclusions, JRC.

37. (2011) Regulation (EU) No 305/2011 of the European Parliament and of the Council of 9 March 2011 laying down harmonised conditions for the marketing of construction products. EUR-Lex, Brussels.

38. (2004) Durable and Reliable Tunnel Structures – The reports (CD Rom). CUR Gouda, The Netherlands.

39. Davidsson G, Haeffler L, Ljundman B, Frantzich H (2003) Handbok för riskanalys (In Swedish). Räddningsverket, Karlstad, Sweden.

40. (2004) Directive 2004/54/EC of the European parliament and of the council on minimum safety requirements for tunnels in the Trans-European Road Network. European Comission, Brussels.

41. EC (2007) Minimum levels of safety in European road tunnels. http://europa.eu/legislation_summaries/transport/road_transport/l24146_en.htm##. Accessed November 8 2012

42. Epstein W (2012) A PRA Practioner Looks at the Fukushima Daiichi Accident. Paper presented at the PSAM11 & ESREL 2012, Helsinki, 25–29 June

43. Epstein W, Yamaguchi A, Laaksonen J, Geller B, Cooke R, Pate-Cornell E, Kitamura M, Kuzmina I, Tappin D, Bot PL (2012) Fukushima – panel discussion. PSAM11 & ESREL 2012. Helsinki

44. Fabbri L, Contini S (2009) Benchmarking on the evaluation of major accident-related risk assessment. J Hazard Mater 162(2–3):1465-1476 http://dx.doi.org/10.1016/j.jhazmat.2008.06.071

45. Fairman R (2007) What Makes Tolerability of Risk Work? Exploring the Limitations of its Aplicability to Othe rRisk Fields. In: Bouder F, Slavin D, Löfstedt RE (eds) The Tolerability of Risk: A New Framework for Risk Management, Earthscan, London. pp 119-136

46. Ferkl L, Dix A (2011) Risk Analysis - from the garden of eden to its seven most deadly sins. Paper presented at the 14th

International Symposium on Aerodynamics and Ventilation of Tunnels (ISAVT 14), Dundee, Scotland, May 11–13

47. Fischhoff B, Kadvany J (2011) Risk: A very short introduction. Oxford University Press, Oxford.

48. Fischhoff B, Lichtenstein S, Slovic P, Derby S, Keeney R (1981) Acceptable Risk. Cambridge University Press, Cambridge.

49. Fischhoff B, Watson S, Hope C (1984) Defining risk. Pol Sci 17(2):123-139 http://dx.doi.org/10.1007/bf00146924

50. Forster C, Kohl B (2012) Ways of improvements in quantitative risk analyses by application of linear evacuation module and interpolation strategies. In: Ingason H, Lönnermark A (eds) Proceedings from the Fifth International Symposium on Tunnel Safety and Security (ISTSS 2012), New York, USA. SP Technical Research Institute of Sweden, pp 627–636

51. Fröderberg M, Thelandersson S (2014) Uncertainty caused variability in preliminary structural design of buildings. Struct Saf 52:183-193 http://dx.doi.org/10.1016/j.strusafe.2014.02.001

52. Funtowicz S, Ravetz J (1990) Uncertainty and Quality in Science for Policy. Kluwer Academic Publishers, Dordrecht, The Netherlands.

53. Funtowicz S, Ravetz J (1992) Three Types of Risk assessment and the Emergence of Post-Normal Science. In: Ka G (ed) Social Theories of Risk, Praeger, Westport, CT, USA. pp 251-274

54. Gandit M, Kouabenan DR, Caroly S (2009) Road-tunnel fires: Risk perception and management strategies among users. Saf Sci 47(1):105-114 http://dx.doi.org/10.1016/j.ssci.2008.01.001

55. Garrick JB (1998) Technological stigmatism, risk perception, and truth. Reliab Eng Syst Saf 59:41-45 http://dx.doi.org/10.1016/S0951-8320(97)00117-8

56. Garrick BJ, Stetkar John W, Bembia Paul J (2010) Quantitative Risk Assessment of the New York State Operated West Valley Radioactive Waste Disposal Area. Risk Anal 30(8):1219-1230

http://dx.doi.org/10.1111/j.1539-6924.2010.01418.x

57. Gehandler J, Wickström U (2014) Estimation of tunnel temperature downstream a tunnel fire considering time dependent wall heat losses. In: Ingason H, Lönnermark A (eds) Proceedings from the Sixth International Symposium on Tunnel Safety and Security (ISTSS 2014), Marseille, France. SP Technical Research Institute of Sweden, pp 195–204

58. Gehandler J, Ingason H, Lönnermark A, Frantzich H (2012) Requirements and verification methods of tunnel safety and design. SP Technical Research Institute of Sweden. Borås, Sweden.

59. Gehandler J, Ingason H, Lönnermark A, Frantzich H, Strömgren M (2013) Performance-based requirements and recommendations for fire safety in road tunnels (FKR-BV12). SP Technical Research Institute of Sweden. Borås, Sweden.

60. Gehandler J, Eymann L, Regeffe M (2014) Limit-based fire hazard model for evaluating tunnel life safety. Fire Technol 50(4):1-30 http://dx.doi.org/10.1007/s10694-014-0406-5

61. Gehandler J, Ingason H, Lönnermark A, Frantzich H, Strömgren M (2014) Performance-based design of road tunnel fire safety: Proposal of new Swedish framework. Case Stud Fire Saf 1(0):18-28 http://dx.doi.org/10.1016/j.csfs.2014.01.002

62. Gildersleeve C, Sherlock W (2014) Do modern fire and life safety standards and codes restrict innovation in urban multi entry and exit road tunnel design and cosntruction? In: Ingason H, Lönnermark A (eds) Proceedings from the Sixth International Symposium on Tunnel Safety and Security (ISTSS 2014), Marseille, France. SP Technical Research Institute of Sweden, pp 299–308

63. Grant G, Jagger S (2012) The use of tunnel ventilation for fire safety. In: Beard A, Carvel R (eds) Handbook of Tunnel Fire Safety, 2nd edn. ICE Publishing, London. pp 177-216

64. Guen J-ML (2007) Applying the HSE's Risk Decision Model: Reducing Risks, Protecting People. In: Bouder F, Slavin D, Löfstedt RE (eds) The Tolerability of Risk: A New Framework

for Risk Management, Earthscan, London. pp 105-118

65. Hadjisophocleous G, Jia Q (2009) Comparison of FDS Prediction of Smoke Movement in a 10-Storey Building with Experimental Data. Fire Technol 45(2):163-177 http://dx.doi.org/10.1007/s10694-008-0075-3

66. Hammond JS, Keeney RL, Raiffa H (1999) Smart choices: a practical guide to making better decisions 2002 edn. Broadway Books, New York.

67. Hansen R, Ingason H (2011) An engineering tool to calculate heat release rates of multiple objects in underground structures. Fire Saf J 46(4):194-203 http://dx.doi.org/10.1016/j.firesaf.2011.02.001

68. Hansen R, Ingason H (2012) Heat release rates of multiple objects at varying distances. Fire Saf J 52(0):1-10 http://dx.doi.org/10.1016/j.firesaf.2012.03.007

69. Hansson SO (2003) Ethical criteria of risk acceptance. Erkenntnis 59(3):291-309 http://dx.doi.org/10.1023/A:1026005915919

70. Hermansson H (2005) Consistent risk management: Three models outlined. J Risk Res 8(7–8):557-568 http://dx.doi.org/10.1080/13669870500085189

71. Hermansson H (2007) A three-party model tool for ethical risk analysis. Risk Management 9(3):129-144 http://dx.doi.org/10.1057/palgrave.rm.8250028

72. Hermansson H (2010) Towards a fair procedure for risk management. J Risk Res 13(4):501-515 http://dx.doi.org/10.1080/13669870903305903

73. Hollnagel E (2006) A function-centred approach to joint driver-vehicle system design. Cognit Tech Work 8(3):169-173 http://dx.doi.org/10.1007/s10111-006-0032-1

74. Hollnagel E (2010) Extending the scope of the human factor. In: Hollnagel E (ed) Safer Complex Industrial Environments, CRC Press, London.

75. Hollnagel E (2011) Prologue: the scope of resilience engineering. In: Hollnagel E, Pariès J, Woods DD, Wreathall

J (eds) Resilience engineering in practice, Ashgate, Farnham, England. pp xxix-xxxix

76. Holm L. Var är visionen? Tidningen Proffs: 2007. http://www.tidningenproffs.se/kronika/2007/10/26/Var-ar-visionen/ (In Swedish). Accessed 15 Jan 2014

77. Holman JP (2010) Heat Transfer. Mc Graw Hill, Boston, USA.

78. Holmgren Å, Thedéen T (2003) Riskanalys. In: Grimvall G, Jacbosson P, Thedéen T (eds) Risker i tekniska system (Swedish), Studentlitteratur, Lund. pp 253-274

79. (2001) Reducing risks, protecting people: HSE's decision-making process. Health and Safety Executive, London.

80. (2010) EN 31010:2010 Risk management - Risk assessment techniques. CENELEC, Brussels.

81. Ingason H (2003) Fire Development in Catastrophic Tunnel Fires (CTF). In: Ingason H (ed) International Symposium on Catastrophic Tunnel Fires (CTF), SP Swedish National Testing and Research Institute, Borås, Sweden. pp 31-47

82. Ingason H (2005) Model Scale Tunnel Fire Tests - Longitudinal ventilation. SP Swedish National Testing and Research Institute. Borås, Sweden.

83. Ingason H (2008) State of the Art of Tunnel Fire Research. Fire Saf Sci 9:33-48 http://dx.doi.org/10.3801/IAFSS.FSS.9-33

84. Ingason H (2012) Fire dynamics in tunnels. In: Beard A, Carvel R (eds) Handbook of Tunnel Fire Safety, 2nd edn. ICE Publishing, London. pp 273-308

85. Ingason H, Li YZ (2010) Model Scale Tunnel Fire Tests- Point extraction ventilation. SP Technical Research Institute of Sweden. Borås, Sweden.

86. Ingason H, Li YZ (2010) Model scale tunnel fire tests with longitudinal ventilation. Fire Saf J 45(6–8):371-384 http://dx.doi.org/10.1016/j.firesaf.2010.07.004

87. Ingason H, Li YZ (2014) Technical trade-offs using fixed fire fighting systems. In: Proceedings from the Seventh International Conference on Tunnel Safety and Ventilation, Graz, Austria.

pp 90–97

88. Ingason H, Lönnermark A (2012) Heat Release Rates in Tunnel Fires: A Summary. In: Beard A, Carvel R (eds) In The Handbook of Tunnel Fire Safety, 2nd edn. ICE Publishing, London. pp 309-328

89. Ingason H, Bergqvist A, Lönnermark A, Frantzich H, Hasselrot K (2005) Räddningsinsatser i vägtunnlar (In Swedish). Räddningsverket, Karlstad, Sweden.

90. Ingason H, Li YZ, Lönnermark A (2015) Tunnel Fire Dynamics. Springer, New York.

91. (2010) Performance-Based Building Regulatory Systems. Inter-Jurisdictional Regulatory Collaboration Committee, Washington.

92. (2006) 16733:2006 Fire safety engineering - Selection of design fire scenarios and design fires. International Organization for Standardization, Geneva.

93. (2009) Risk management - Vocabulary. International Organization for Standardization, Geneva.

94. (2009) 2009(E) Fire-safety engineering: Technical information on methods for evaluating behaviour and movement of people. International Organization for Standardization, Geneva.

95. (2009) 23932:2009 Fire safety engineering - General principles. International Organization for Standardization, Geneva.

96. (2012) 13571 :2012 (E) Life threatening components of fire -- Guidelines for the estimation of time to compromised tenability in fires. International Organization for Standardization, Geneva.

97. (2012) 16732–1 Fire safety engineering - Fire risk assessment - Part 1: General. International Organization for Standardization, Geneva.

98. Johannesson P, Bergman B, Svensson T, Arvidsson M, Lönnqvist Å, Barone S, de Maré J (2013) A Robustness Approach to Reliability. Qual Reliab Eng Int 29:17-32 http://

dx.doi.org/10.1002/qre.1294

99. Johansson H (2001) Decision Making in Fire Risk Management. Lunds Universitet, Lund.

100. Johansson P, Chakhunashvili A, Barone S, Bergman B (2006) Variation Mode and Effect Analysis: a Practical Tool for Quality Improvement. Qual Reliab Eng Int 22:865-876 http://dx.doi.org/10.1002/qre.773

101. Johnson P (2012) Fire Safety Engineering: A Tool in Tunnel Design. In: Ingason H, Lönnermark A (eds) Proceedings from the Fifth International Symposium on Tunnel Safety and Security (ISTSS 2012), New York, USA. SP Technical Research Institute of Sweden, pp 57–68

102. Kaplan S (1991) The general theory of quantitative risk assessment. In: Haimes YY, Moser DA, Stakhiv EZ (eds) Risk-Based Decision Making in Water Resources V, British Library, Santa Barbara, California, United States. pp 11-39

103. Kaplan S (1997) The Words of Risk Analysis. Risk Anal 17(4):407-417 http://dx.doi.org/10.1111/j.1539-6924.1997.tb00881.x

104. Kaplan S, Garrick JB (1981) On the quantitative definittion of Risk. Risk Anal 1(1):11-27 http://dx.doi.org/10.1111/j.1539-6924.1981.tb01350.x

105. Karlsson B, Quintiere JG (1999) Enclosure fire dynamics. CRC Press, London.

106. Kazaras K, Kirytopoulos K, Rentizelas A (2012) Introducing the STAMP method in road tunnel safety assessment. Saf Sci 50(9):1806-1817 http://dx.doi.org/10.1016/j.ssci.2012.04.013

107. Kim HK, Lönnermark A, Ingason H (2010) Effective fire fighting operations in road tunnels. Sweden, SP, Borås.

108. Kirytopoulos K, Kazaras K (2011) The need for a new approach in road tunnels risk analysis. In: Soares CG (ed) ESREL 2011, Troyes French. CRC Press, pp 2562–2569. http://dx.doi.org/10.1201/b11433-363

109. Kuligowski ED, Peacock RD, Hoskins BL (2010) A Review of Building Evacuation Models, 2nd edition. NIST, Frie Research Division, Technical Note 1680

110. Latané B, Darley L (1970) The unresponsive bystander: Why doesn't he help?. Meredith Corporation, New York.

111. Lauridsen K, Christou M, Amendola A, Markert F, Kozine I (2001) Assessing the uncertainties in the process of risk analysis of chemical establishements: Part II. In: Zio E, Demichela M, Piccinini N (eds) Towards a Safer World - Proceedings of the ESREL Conference, Turin, Italy. pp 16-20

112. Lauridsen K, Christou M, Amendola A, Markert F, Kozine I, Fiori M (2001b) Assessing the uncertainties in the process of risk analysis of chemical establishements: Part I. In: Zio E, Demichela M, Piccinini N (eds) Towards a Safer World - Proceedings of the ESREL Conference, Turin, Italy. pp 599–606

113. Lauridsen K, Kozine I, Markert F, Amendola A, Christou M, Fiori M (2002) Assessment of Uncertainties in Risk Analysis of Chemical Establishments: Final summary report. The ASSURANCE project, Risoe National Laboratory, Roskilde, Denmark.

114. Lille GH, Andersen T (1996) Acceptance of risks related to the transport of dangerous goods through road tunnels. In: OECD-ERS2 working group - Seminar on decision models for the tranportation of dangerous goods through road tunnels, Oslo, Norway.

115. Lönnermark A (2007) Goods on HGVs during Fires in Tunnels. In: 4th International Conference on Traffic and Safety in Road Tunnels, Hamburg, Germany.

116. Lönnermark A, Ingason H (2007) The Effect of Cross-sectional Area and Air Velocity on the Conditions in a Tunnel during a Fire. SP Report 2007:05. SP Technical Research Institute of Sweden, Borås, Sweden.

117. Lutz RR (1993) Analyzing software requirements errors in safety-critical, embedded systems. In: Proceedings of IEEE

International Symposium on Requirements Engineering. SP Technical Research Institute of Sweden, Pasadena, CA, USA. pp 126-133

118. Malmtorp J, Vedin P (2014) An alternative approach to safety in road tunnels. In: Ingason H, Lönnermark A (eds) Proceedings from the Sixth International Symposium on Tunnel Safety and Security (ISTSS 2014), Marseille, France. SP Technical Research Institute of Sweden, pp 309–316

119. Martens MH, Jenssen GD (2012) Human behaviour in tunnels what further steps to take? In: Ingason H, Lönnermark A (eds) Proceedings from the Fifth International Symposium on Tunnel Safety and Security (ISTSS 2012), New York, USA. SP Technical Research Institute of Sweden, pp 69–86

120. Mawhinney J (2011) Fixed Fire Protection Systems in Tunnels: Issues and Directions. Fire Technol 49(2):477-508 http://dx.doi.org/10.1007/s10694-011-0220-2

121. Meacham BJ (2004) Decision-Making for Fire Risk Problems: a Review of Challenges and Tools. J Fire Protect Eng 14(2):149-168 http://dx.doi.org/10.1177/1042391504040262

122. Meacham BJ (2004) Understanding risk: Quantification, perceptions, and characterization. J Fire Protect Eng 14(3):199-227 http://dx.doi.org/10.1177/1042391504042454

123. Meacham BJ, Custer RLP (1995) Performance-Based Fire Safety Engineering: An Introduction of Basic Concepts. J Fire Protect Eng 7(2):35-53 http://dx.doi.org/10.1177/104239159500700201

124. Menzies J (2007) Creating a culture of safety: Private fleets share tips on implementing safety programs. Transportation and logistics, Canadian.

125. Möller N (2006) Safety and decision-making. Stockholm, Royal Institute of Technology (KTH).

126. Möller N (2009) Should we follow the experts' advice? Epistemic uncertainty, consequence dominance and the knowledge asymmetry of safety. Int J Risk Assess Manag 11(3–4):219-236 http://dx.doi.org/10.1504/IJRAM.2009.023154

127. Möller N, Hansson SO (2008) Principles of engineering

safety: Risk and uncertainty reduction. Reliab Eng Syst Saf 93(6):798-805 http://dx.doi.org/10.1016/j.ress.2007.03.031

128. Mooney GH (1977) The valuation of human life. Macmillan, London.

129. Morgan MG, Henrion M (1990) Uncertainty. Cambridge University Press, New York.

130. Nævestad T-O, Meyer S (2014) A survey of vehicle fires in Norwegian road tunnels 2008–2011. Tunnelling and Underground Space Technology 41(0):104-112 http://dx.doi.org/10.1016/j.tust.2013.12.001

131. Nilsen AR, Log T (2009) Results from three models compared to full-scale tunnel fires tests. Fire Saf J 44(1):33-49 http://dx.doi.org/10.1016/j.firesaf.2008.03.001

132. Nilsson D (2009) Exit choice in fire emergencies - Influencing choice of exit with flashing lights. Lund University, Lund.

133. Nilsson D, Johansson A (2009) Social influence during the initial phase of a fire evacuation - Analysis of evacuation experiments in a cinema theatre. Fire Saf J 44(1):71-79 http://dx.doi.org/10.1016/j.firesaf.2008.03.008

134. Noizet A (2012) Egress behaviour during road tunnel fires. In: Carvel R, Beard A (eds) Handbook of Tunnel Fire Safety, 2nd edn. ICE Publishing, London. pp 421-438

135. (1982) Risk and Decision Making: Perspectives and Research. National Research Council, Washington, DC.

136. Nývlt O, Prívara S, Ferkl L (2011) Probabilistic risk assessment of highway tunnels. Tunnelling and Underground Space Technology 26(1):71-82 http://dx.doi.org/10.1016/j.tust.2010.06.010

137. Oggero A, Darbra RM, Muñoz M, Planas E, Casal J (2006) A survey of accidents occurring during the transport of hazardous substances by road and rail. J Hazard Mater 133(1–3):1-7 http://dx.doi.org/10.1016/j.jhazmat.2005.05.053

138. Oppenheim I, Shinar D (2012) A context-sensitive model of driving behaviour and its implications for in-vehicle safety

systems. Cogn Tech Work 14(3):261-281 http://dx.doi.org/10.1007/s10111-011-0178-3

139. Otway H (1992) Public Wisdom, Expert Fallability: Toward a Contextual Theory on Risk. In: Ka G (ed) Social Theories of Risk, Praeger, Westport, CT, USA. pp 215-228

140. Paté-Cornell ME (1996) Uncertainties in risk analysis: Six levels of treatment. Reliab Eng Syst Saf 54(2–3):95-111 http://dx.doi.org/10.1016/S0951-8320(96)00067-1

141. (2007) Integrated approach to road tunnel safety (2007R07). World Road Association. La Défense cedex, France.

142. (2008) Risk analysis for road tunnels. World Road Association, La Défense cedex, France.

143. Pidgeon N (1997) The Limits to Safety? Culture, Politics, Learning and Man–Made Disasters. J Contingencies and Crisis Management 5(1):1 http://dx.doi.org/10.1111/1468-5973.00032

144. Proulx G, Sime J (1991) To Prevent 'Panic' In An Underground Emergency: Why Not Tell People The Truth? Fire Saf Sci 3:843-852 http://dx.doi.org/10.3801/IAFSS.FSS.3-843

145. Purser D (2009) Hazards from toxicity and heat in fires. Hatford Environmental Research, Hatford.

146. Rasmussen J, Svedung I (2000) Proactive risk management in a dynamic society. Räddningsverket (Swedish rescue service agency), Karlstad, Sweden

147. Rattei G, Lentz A, Kohl B (2014) How frequent are fire in tunnels - Analysis from Austrian tunnel incident statistics. In: Proceedings from the Seventh International Conference on Tunnel Safety and Ventilation, Graz, Austria. pp 5–11

148. Reason J (1990) Human Error. Cambridge University Press, Cambridge.

149. Reason J (1997) Managing the risks of organizational accidents. Ashgate, Farnham, England.

150. Rein G, Torero JL, Jahn W, Stern-Gottfried J, Ryder NL, Desanghere S, Lázaro M, Mowrer F, Coles A, Joyeux D,

Alvear D, Capote JA, Jowsey A, Abecassis-Empis C, Reszka P (2009) Round-robin study of a priori modelling predictions of the Dalmarnock Fire Test One. Fire Saf J 44(4):590-602 http://dx.doi.org/10.1016/j.firesaf.2008.12.008

151. Renn O (1998) Three decades of risk research: accomplishments and new challenges. J Risk Res 1(1):49-71 http://dx.doi.org/10.1080/136698798377321

152. Renn O (2008) Risk gouvernance: coping with uncertainty in a complex world. Earthscan, London.

153. Rosmuller N, Beroggi GEG (2004) Group decision making in infrastructure safety planning. Saf Sci 42(4):325-349 http://dx.doi.org/10.1016/S0925-7535(03)00046-8

154. Ruijter HA (2012) Safety of Dutch tunnels guaranteed by standard approach. In: Ingason H, Lönnermark A (eds) Proceedings from the Fifth International Symposium on Tunnel Safety and Security (ISTSS 2012), New York, USA. SP Technical Research Institute of Sweden, pp 283–288

155. Ruland T, Snel A (2010) Determination and analysis of tunnel safety requirements from a functional point of view. In: Ingason H, Lönnermark A (eds) Proceedings from the Fourth International Symposium on Tunnel Safety and Security (ISTSS), Frankfurt, Germany. SP Technical Research Institute of Sweden, pp 557–560

156. Ruland T, Daverveld T, Duijnhoven Bv, Gelder Jv, Krouwel R, Teeuw J-M (2012) An integrated functional design approach for safety related tunnel processes. In: Ingason H, Lönnermark A (eds) Proceedings from the Fifth International Symposium on Tunnel Safety and Security (ISTSS 2012), New York, USA. SP Technical Research Institute of Sweden, pp 167–176

157. Santos-Reyes J, Beard AN (2003) A systemic approach to safety management on the british railway system. Civ Eng Environ Syst 20(1):1-21 http://dx.doi.org/10.1080/10286600302232

158. Santos-Reyes J, Beard AN (2006) A systemic analysis of the Paddington railway accident. Proceedings of the Institution of Mechanical Engineers, Part F. J Rail and Rapid Transit

220(2):121-151 http://dx.doi.org/10.1243/09544097JRRT33

159. Santos-Reyes J, Beard AN (2011). A preliminary analysis of the 1996 Channel Tunnel fire. In: 3rd International Tunnel Safety Forum for Road and Rail, Nice, France.

160. Santos-Reyes J, Beard A (2012) A systemic approach to tunnel fire safety management. In: Carvel R, Beard A (eds) Handbook of Tunnel Fire Safety, 2nd edn. ICE Publishing, London. pp 485-508

161. Shields J (2012) Human behaviour during tunnel fires. In: Carvel R, Beard A (eds) Handbook of Tunnel Fire Safety, 2nd edn. ICE Publishing, London. pp 399-420

162. Sime J (1985) Movement toward the Familiar Person and Place Affiliation in a Fire Entrapment Setting. Environ Behav 17(6):697-724 http://dx.doi.org/10.1177/0013916585176003

163. Sime J, Creed C, Kimura M, Powell J (1992) Human behaviour in fires. Joint Committee on Fire Research, London.

164. Sleich JB, Cajot LG, Pierre M (2002) Competitive steel buildings through natural fire safety concepts. Directorate-General for Research and Innovation Luxembourg, European Commission.

165. Slovic P (1987) Perception of Risk. Science 236(4799):280-285 http://dx.doi.org/10.1126/science.3563507

166. Slovic P (2000) The Perception of Risk. Routledge, London.

167. Stamatelatos M, Apostolakis G, Dezfuli H, Guarro S, Moieni P, Mosleh A, Paulos T, Youngblood R (2002) Probabilistic risk assessment proceedure guide for NASA managers and Practitioners. NASA Office of Safety and Mission Assurance, Washington, DC.

168. Stamatelatos M, Vesely W, Dugan J, Fragola J, III JM, Railsback J (2002) Fault tree handbook with aerospace aplications. NASA Office of Safety and Mission Assurance, Washington, DC.

169. Svensson T, Johannesson P (2013) Reliable fatigue design, by rigid rules, by magic, or by enlightened engineering. In: 5th

Fatigue Design Conference, Fatigue Design, Senlis, France.

170. Thomas P (1958) The movement of buoyant fluid against a stream and the venting of underground fires. Fire Research Note 351. Fire Research Station, Borehamwood.

171. Thomas PH (1968) The Movement of Smoke in Horizontal Passages Against an Air Flow. Fire Research Note 723. Fire Research Station, Borehamwood.

172. Thomas P (1986) Design guide: Structure fire safety CIB W14 Workshop report. Fire Saf J 10(2):77-137 http://dx.doi.org/10.1016/0379-7112(86)90041-X

173. Tong D, Canter D (1985) The decision to evacuate: a study of the motivations which contribute to evacuation in the event of fire. Fire Saf J 9(3):257-265 http://dx.doi.org/10.1016/0379-7112(85)90036-0

174. UPI. Fire in Norwegian tunnel hospitalizes 70. 2013. upi.com: http://www.upi.com/Top_News/World-News/2013/08/05/Fire-in-Norwegian-tunnel-hospitalizes-70/88231375728950/. Accessed 15 Jan 2014

175. Vaitkevicius A, Colella F, Carvel R (2014) Rediscovering the Throttling Effect. In: Ingason H, Lönnermark A (eds) Proceedings from the Sixth International Symposium on Tunnel Safety and Security (ISTSS 2014), Marseille, France. SP Technical Research Institute of Sweden, pp 373–378

176. Vareman N, Persson J (2010) Why separate risk assessors and risk managers? Further external values affecting the risk assessor qua risk assessor. J Risk Res 13(5):687-700 http://dx.doi.org/10.1080/13669871003660759

177. Vrijling JK, van Hengel W, Houben RJ (1995) A framework for risk evaluation. J Hazard Mater 43(3):245-261 http://dx.doi.org/10.1016/0304-3894(95)91197-V

178. Vrijling JK, van Hengel W, Houben RJ (1998) Acceptable risk as a basis for design. Reliab Eng Syst Saf 59(1):141-150 http://dx.doi.org/10.1016/S0951-8320(97)00135-X

179. Weerheijm J (2014) Berg Bvd Explosion risks and consequences for tunnels. In: Ingason H, Lönnermark A (eds) Proceedings

from the Sixth International Symposium on Tunnel Safety and Security (ISTSS 2014), Marseille, France. SP Technical Research Institute of Sweden, pp 46–61

180. Wynne B (1992) Risk and Social Learning: Reification to engagement. In: Ka G (ed) Social Theories of Risk, Praeger, Westport, CT, USA. pp 275-297

Chapter 5

Integrative Approach to the Plant Commissioning Process

Kris Lawry[1] and Dirk John Pons[2]

[1]Department of Chemical and Process Engineering, University of Canterbury, Private Bag 4800, Christchurch 8020, New Zealand

[2]Department of Mechanical Engineering, University of Canterbury, Private Bag 4800, Christchurch 8020, New Zealand

ABSTRACT

Commissioning is essential in plant-modification projects, yet tends to be ad hoc. The issue is not so much ignorance as lack of systematic approaches. This paper presents a structured model wherein commissioning is systematically integrated with risk management, project management, and production engineering. Three strategies for commissioning emerge, identified as direct, advanced, and parallel. Direct commissioning is the traditional approach of stopping the plant to insert the new unit. Advanced commissioning is the commissioning of the new unit prior to installation. Parallel commissioning is the commissioning of the new unit in its operating position, while the old unit is still operational. Results are reported for two plant case studies, showing that advanced and parallel commissioning can significantly reduce risk. The model presents a novel and more structured way of thinking about commissioning, allowing for a more critical examination of how to approach a particular project.

INTRODUCTION

Background

Plant modifications are an ongoing process throughout the life of any process plant. Reasons for modification include efforts to improve reliability, production capacity, quality, or productivity. Seamless incorporation is the key concern associated with the installation of any new equipment in an operating plant due to the high cost of process downtime. Several steps can be taken to minimise the risk associated with the installation of new equipment such as hazard and operability studies, project management, development of redundancy plans, and commissioning of the new equipment.

Of these, commissioning is an essential activity in many plant-modification projects and has significant implications for project success. Yet paradoxically it tends to be approached in an ad hoc manner. It is often included in project plans, so it is not that people are ignorant of commissioning. Rather, the problem is that there is a lack of systematic approaches to commissioning, so it is frequently left to tradespeople and plant operators to manage in whatever way they see fit. This is an undesirable situation since it results in unpredictable outcomes. In some cases it can even cause serious problems. An extreme example would be the catastrophic failure of the Chernobyl nuclear power plant (1986), which was caused by operators attempting an ad hoc test of the efficacy of a modified emergency cooling system.

This paper presents a structured conceptual model for the commissioning process, and two cases studies showing application to operating plant.

EXISTING MODELS OF COMMIS-SIONING

Literature

Many authors have highlighted the value of commissioning from a range of different perspectives but they all agree that commissioning and the integration of a new project is critical to the success of any project [1–10]. However commissioning is poorly defined and is interpreted ambiguously [6, 11], which leads to inefficient utilisation within industry. In this paper "commissioning" is defined as the disciplined activity involving careful testing, calibration, and proving of all systems, software, and networks within the project boundary [5].

Current Models of Commissioning

Factors that are known to affect the commissioning process include the following.

- Type of project. Thus situational variables are important; that is, the factors that resulted in a successful (or failed) commissioning outcome in one case are not necessarily transferrable to a different situation.
- Who is in charge of the phase. Commissioning can be completed by a range of different groups depending on the project. It can be the equipment manufacturers, operation

team, or a separate commissioning team depending on scale and requirements of the project [12]. The relationships between these people are also important (social dimension) [6, 13], hence also contractual obligations (see (iv) below).

- Number/type of phases. Commissioning can also be broken down in several sections such as planning, precommissioning, testing, integration, monitoring, documenting, and handover depending on the level of complexity of the project. This requires careful project planning (see (iv) below).

- Project planning and contractual sufficiency. It is widely recognised in the literature that commissioning requires deliberate planning, as opposed to ad hoc treatment. Thus it needs appropriate consideration in the work breakdown structure and project planning [14], allocation of resources, transferral of those costs into the initial contract [9, 15–18], and creation of specific operating procedures (especially important for safety-critical plant like boilers [19]). This corresponds to the "integration" tasks in the project management approach [6].

The commissioning process has been examined for a wide range of different projects [2–5, 8, 20, 21]. The predominate approach can be described as task specific; the literature tends to identify specific tasks that should be completed as part of commissioning. Thus the focus has been on completing multiple checks on a system to ensure it will operate as expected. Thus there are many reports in the literature, too numerous to mention, about commissioning experiences in specific case studies. These are undoubtedly helpful, especially for lessons learned and application to comparable situations. They are also systematic, in a way, especially in the provision of templates and checklists to guide practitioners.

However there is a lack of holistic or integrative models. There is much less literature at the next higher level of abstraction, which is the commissioning process in general. At this level we are interested not so much in case-specific experiences but in the fundamental principles and the methodology. What exists at this level is primarily in the area of instrumentation and control; some examples are [5, 22, 23]. Thus the existing commissioning strategies in the literature can be categorised into three types, see Figure 1. These are (a) ad hoc, which is action-orientated problem solving; and (b) template, which involves using a checklist, or operating procedure that worked before or in another situation. Both (a) and (b) are premised on the assumption that commissioning is a routine set of tasks. The third strategy challenges that premise and calls for a deliberately thoughtful approach. Thus the third category in the literature is (c) methodological, which involves analysis of the situational needs and deliberate selection of the most relevant of several possible commissioning methods.

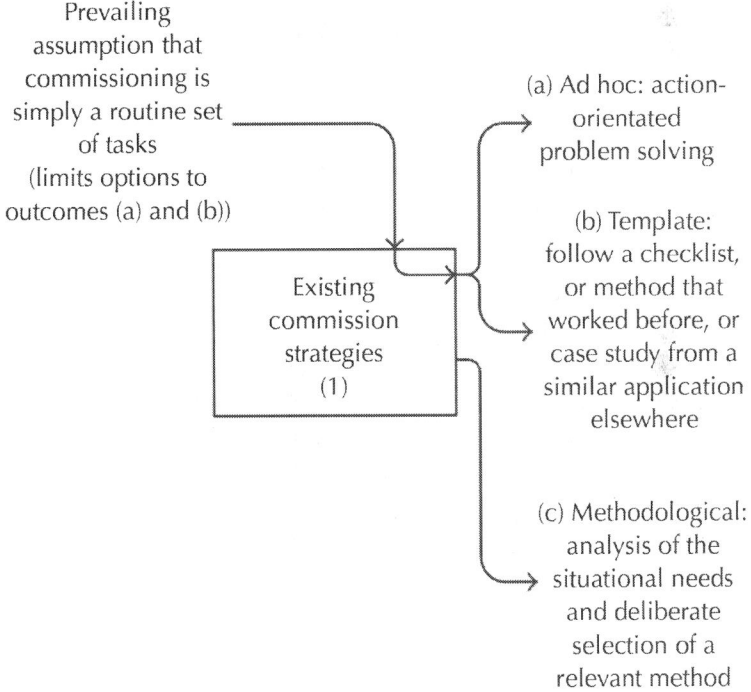

Figure 1: Three common strategies for commissioning, broadly reflecting the approaches described in the literature. If commissioning is perceived to be simply a routine set of tasks, which is a common assumption, then this tends to preclude any more thoughtful approach to the problem.

Issues and Problem Areas

A clear refrain in the literature is that commissioning (i) needs deliberate project management, but (ii) is too often not given the attention it deserves. One of the issues with commissioning, which contributes to problem (ii), is that the value thereof is hard to quantify. Justifying the value of commissioning may be completed using qualitative analysis similar to quantification of risk in a project [24]. This is based on the consequence and probability of the system failing to operate as anticipated. In other cases there is no attempt at justification at all, so the value is not appreciated.

Another issue is the tendency to underresource the commissioning in the project planning, which is issue (i) above. Underresourcing is due to several factors such as its omission in the project management. There is often a high level of variability as a result of the case-specific nature making it difficult to fit into the established planning structure. Existing project management frameworks, such as the PMBOK [9], are general approaches. While they acknowledge the commissioning stage they do not, and cannot reasonably be expected to, provide case-specific guidance on commissioning. They treat commissioning very lightly and rely on the practitioner to identify whether or not commissioning is an important part of the project. The literature suggests that practitioners too often fail to realise the importance and therefore fail to plan sufficiently. Alternatively, project managers may simply be overly optimistic about the risks associated with the installation of a new system. Whatever the reason, the result can be insufficient resources being allocated, with the consequence of poor completion. Incorporation of a broad conceptual model of commissioning into the project management practices would be the first logical step. Commissioning draws from several project knowledge areas such as integration, communication, and risk management. The logical approach is to incorporate into the project life cycle between the execution and closing phases [4, 6, 8].

Problem Definition

Current models of commissioning tend to be simplistic, and relevant only to specific areas. They are focused on the process and consequently tend to produce a somewhat prescriptive list of tasks that need to be performed. A higher-level reconceptualisation of the commissioning process, with the development of a more general theory, could be valuable.

The purpose of this work is to develop a more holistic and integrative theory of commissioning. The specific emphasis is on reducing process downtime, without compromising plant reliability. This is worth attempting as it has the potential to provide a general framework in which the other more process-specific models can be placed.

APPROACH

We start by reconceptualising commissioning in broad terms. We categorise the commissioning strategies according to the operational risk. This results in three categories: direct, parallel, and advanced. We then apply a system modelling method to embed these within the broader manufacturing context. Finally we apply the new framework to two case studies to demonstrate the applicability.

RESULTS

Categorisation of Commissioning Projects

Starting Premise

We start with the premise that the value of commissioning is essentially one of systematic risk reduction, that is, used to minimise the risk associated with the installation of a new piece of equipment. More specifically the application of commissioning for the installation of new equipment into the process industry reduces the risk of equipment damage, environmental health and safety, and process downtime.

Thus commissioning is a strategy for treating risk [24]. This has the further important implication that the treatment, hence type of commissioning, can be aligned with the degree of technical risk that the organisation can accept. Thus we specifically link commissioning, as a treatment strategy, to the concept of "tolerable risk" within the risk management literature, and to the concept of strategic risk for the organisation as a whole [25]. This also has contractual implications in project-setup phase, where there is a need to differentiate between the commissioning risk elements and proportion them between the equipment manufacturer, project management organisation, and plant owner [26].

From this starting assumption we identify three categories of commissioning, as strategies in response to organisational risk-tolerance. These are direct commissioning, advanced commissioning, and parallel commissioning. Each has strengths and weaknesses. They can be deployed individually or together.

Direct Commissioning

Direct commissioning is the classical approach to commissioning where the new equipment is installed and the system must remain offline as commissioning is completed. Direct commissioning is the most straightforward approach as no additional equipment or simulation is required. The new equipment is installed into its operational position and the process cannot restart until the system has been commissioned and is running correctly. There is a high level of downtime in this process as the whole system cannot be operated until the new unit is electrically, mechanically, and operationally tested. There is also the risk of having to reinstall the old unit if there are significant complications at any phase of the commissioning process. Direct commissioning is often reserved for well-established unit operations such as new pumps and flow meters. Direct commissioning is most effective when it is used on well-established system and ones that are not a key requirement of the process.

Advanced Commissioning

Advanced commissioning is the process of operating the new unit in advance of installation and in isolation of the main process operation. Advanced commissioning requires the simulation of all proprietary systems that interact with the new unit. Simulation can be extremely complicated or simple depending on the level of interaction between the process and the new unit. (In this context "simulation" can refer to the artificial provision of physical inputs to the new machine or unit, smaller scale models, and mathematical modelling of the functional behaviour of the unit.) Advanced commissioning allows for the electrical, mechanical, and part of the operational testing to be completed. The full functionality of the unit cannot be proven as the system is being simulated by external means, which will always be an approximation of reality. Advanced process is extremely valuable for the development of new technology as it allows for the verification of novel processes at low risk. The most common type of advanced commissioning is the development of model systems which both simulate the operation of the system and the new unit. Advanced commissioning can also include computer simulation of new process which provides a cost effective method of developing concepts in the early stages of design. Advanced commissioning is valuable at proving conceptual designs of new technology. The main drawback of advanced commissioning is that the process is only simulated so there is still the potential that the unit can fail when installed into its operational environment.

Parallel Commissioning

Parallel commissioning is the testing of the new system in parallel to the operating system. Parallel commissioning is the most rigorous form of physical and operational commissioning. It allows for the new unit to be tested under full operational conditions, with low risk of significant process interruptions due to the added redundancy of the old system present in an operational capacity. However it also has the highest cost as it requires the duplicate hardware systems and additional structural space. The only risk associated with parallel commissioning is the integration between the two systems. Often there is some type of switching or merging component in these systems which may require minor process stoppage for installation. Parallel commissioning is often completed when it is critical that the process must not stop for any extended period of time. It often lends itself to processes with few interactions between new unit operation and the rest of the process. Parallel commissioning is seldom utilised due to the requirement of a process that can accommodate both the new and old unit.

Conceptual Model

Having identified three types of commissioning, we next seek to set those within a conceptual framework. This is worth doing as it has the potential to identify the situational variables relevant to each type of commissioning. This in turn can be used to further build a theoretical foundation, and provide guidance to practitioners.

Approach

The modelling method uses a structured, deductive process to decompose the process being analysed into multiple subactivities (functions) and for each deduce the initiating events, the controls that determine the extent of the outputs, the inputs required, the process mechanisms that are presumed to support the action, and the outputs. The model was then inductively reconciled with elements of the existing body of knowledge on this topic, and successively refined. The end result is a graphical model that describes the relationships between variables, thereby providing a synthesis of what is known and surmised about the topic. The model is expressed as a series of flowcharts using the integration definition zero (IDEF0) notation [27, 28]. With IDEF0 the object types are inputs, controls, outputs, and mechanisms (ICOM) and are distinguished by placement relative to the box, with inputs always entering on the left, controls above, outputs on the right, and mechanisms below.

Develop Production Capability (Prd-1)

The broader context is that commissioning occurs as part of the development of production capability, and our model starts at this level. (This is already the second level into the model; the top level, which is not shown here, includes product design, operation of the plant, control of production flow, quality, distribution to market, packaging, health and safety, lean/JIT, among other activities. However the present paper focuses on the commissioning activities.) See Figure 2. Commissioning is included as element 5 and occurs towards the end of the plant-development process. Other important activities are the following.

- Determine manufacturing/production sequence.
- Design of the production plant, which also includes the plant layout, material handling, plant control and automation, and (for manufacturing) the development of production tooling and flow control, for example, just-in-time (JIT). Analysis of technology risk (9) is another activity associated with the design phase.
- Building the production system (4), and the associated project management activities.
- Decommissioning the plant (7).
- Project management (8). We note the importance of project management methods in supporting many of the activities of commissioning. There are several models of project

management that might be inserted here, including [9, 29], but these are not specific to commissioning and therefore not detailed further at this point.

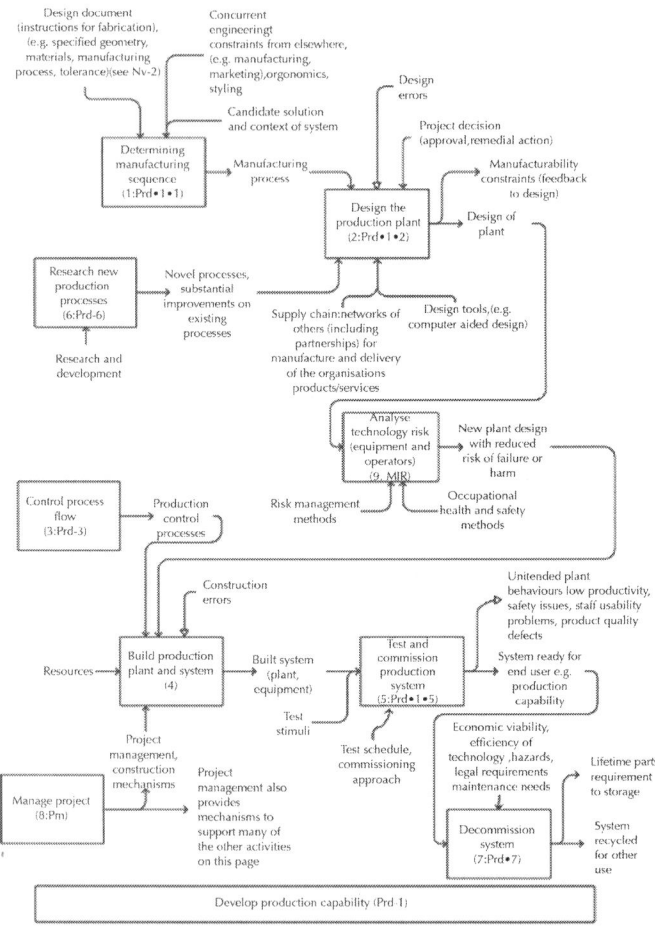

Figure 2: Model for the development of production capability.

We do not deal with these other activities here, but instead move the focus to the test and commission activities. Before doing so, we draw attention to some hollow arrows, which represent errors, in particular design and construction errors, at (2) and (4), respectively, and the possibility for unintended plant behaviours at

(5): low productivity, safety issues, staff usability problems, product quality defects, and so forth. This point is important because the commissioning model that follows specifically seeks to address these risks.

Test and Commission Production System (Prd-1-5)

The model for commissioning a new piece of plant equipment is shown in Figure 3 (Prd-1-5). The conventional commissioning process is included here, as are the new concepts for commissioning approach. One of the conventional activities is to verify instrumentation and control systems (1), which involves the systematic checking of installed hardware against plant schematics. The checks are progressively done for connectivity, cold operation, hot operation, and process control. We do not detail those processes here and instead refer the reader to source material [5] which has information that is useful to practitioners. The final objective of commissioning is also well known, to deliver an operational production system (6) that is ready for the client to use.

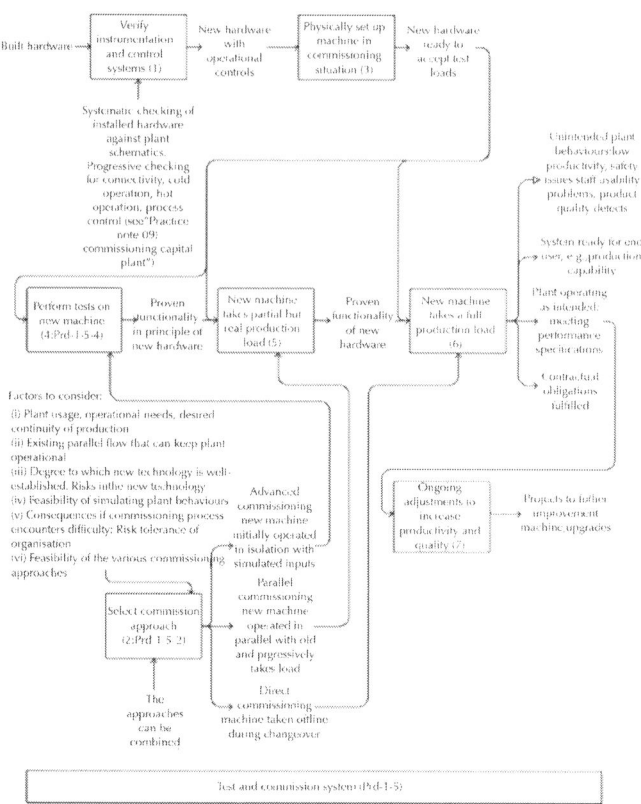

Figure 3: Model for the test and commissioning activities.

Where our model differs is the inclusion of a deliberate stage of deciding which of three commissioning approaches to use in the situation (2): direct, advanced, or parallel. We also note in passing that the quality and lean imperative for continuous improvement will generally mean that there will be ongoing adjustments to increase productivity and quality (7) after the machine has been commissioned. Thus commissioning the machine and closing the contract with the client may be the end of the involvement of the machine builders, but are not the end of the life cycle for the machine itself. This again has contractual implications in the form of service and warranty support from the vendors, and reliability centred maintenance by the plant operators. There is also the decommissioning to consider, which can be a project in itself. (In extreme cases, e.g., nuclear power plant, the cost of decommissioning is comparable to the initial construction cost. If there has been a catastrophic failure of the plant then the decommissioning cost can vastly exceed the construction cost.)

Select Commission Approach (Prd-1-5-2)

The decision involves a choice of direct, advanced, or parallel commissioning. These are not mutually exclusive. Instead some of them may be done sequentially, as shown in Figure 4. For example, advanced commission may precede either of the other two.

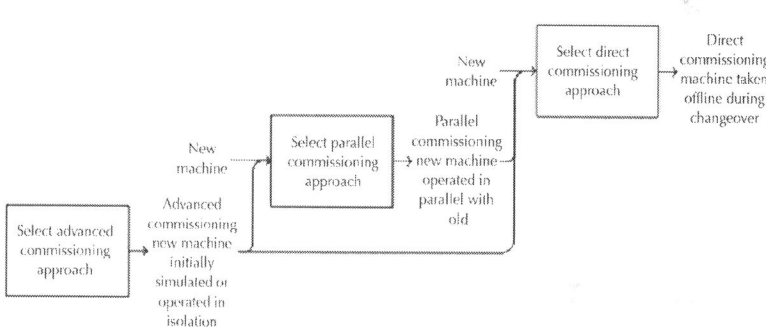

Figure 4: Relationship between the three commissioning approaches: advanced, parallel, and direct.

The various factors relevant to this commissioning decision are anticipated and clustered in groups: ability to recover from a failed installation, assessed or perceived technology risk, desired operational continuity, and timing considerations. The detailed model and the factors within each cluster are shown in Figure 5.

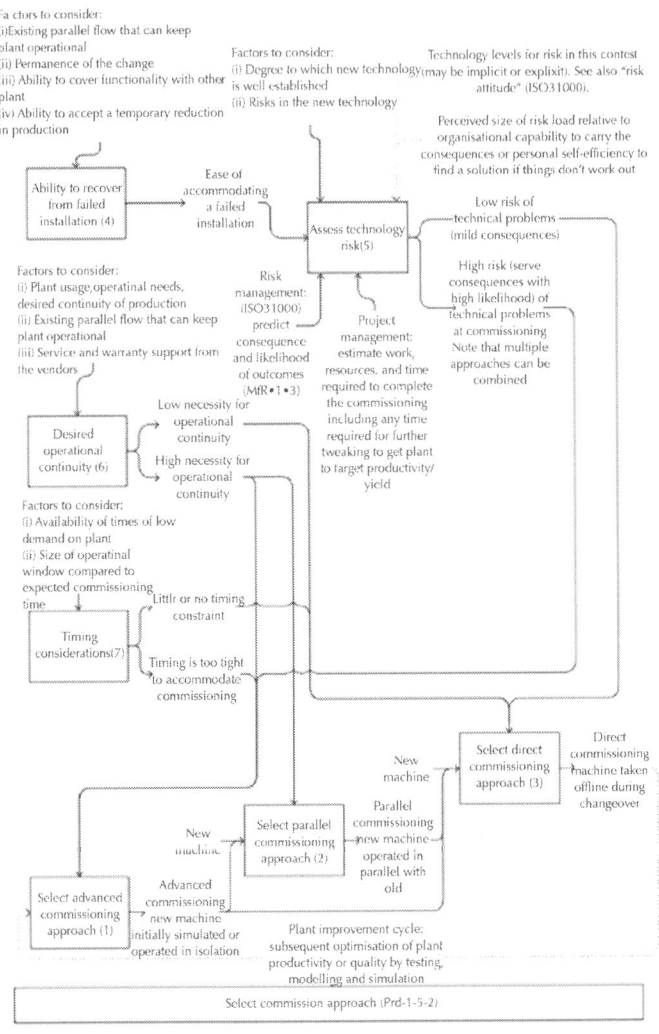

Figure 5: Factors relevant to the commissioning decision.

At this stage the model is primarily logical and qualitative and is intended as a debiasing tool and a guide to action rather than a decision algorithm. It is also a framework for further research in that it proposes subjective relationships of causality that can subsequently be tested and developed as appropriate. (It may even be that in certain areas it could be possible to develop a mathematical model to support the decision, particularly in well-defined areas. Specifically, the model incorporates risk assessment and it is not impossible that there could be well-defined situations where the variables can be determined with sufficient precision that a quantitative risk assessment coupled with (say) a Boolean consideration of the other factors might make for a sufficient mathematical model. However further research would be required to take it to this level of detail.)

Thus the model proposes that the following decision factors are relevant.

- Advanced commissioning is appropriate where technology risk is high, operational continuity is required, and timing constraints are tight.
- Parallel commissioning is appropriate where operational continuity is required and timing constraints are tight.
- Direct commissioning is appropriate where technology risk is low, operational continuity can be disrupted, and timing constraints are loose.

Finally, to complete this part of the conceptual framework, a model is provided for the testing activities of commissioning; see Figure 6.

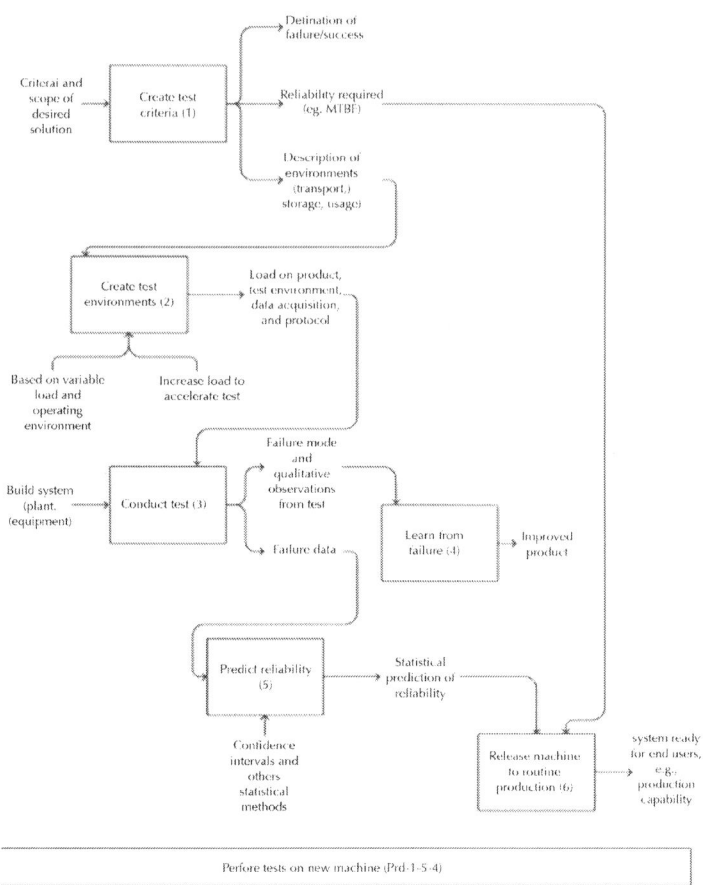

Figure 6: Model for the test activities.

CASE STUDIES

Two cases studies were completed to determine the relevance of this commissioning model in the process industry. First was the development of a novel vertical screw system in the fertilizer industry which used the advanced approach to commissioning. Second was the installation of a ship unloader in the aluminium industry which used a parallel approach to commissioning.

Vertical Screw Project

Ballance Agri-Nutrients single superphosphate plant at Awarua (New Zealand) had recently designed and installed a new phosphate rock feed system. A new vertical screw system was developed to replace the old gravity feed vertical chute which was prone to blockages in the highly reactive and humid environment present in the reaction chamber. The vertical screw was designed to increase the reliability of the process by forcing the rock into the reaction chamber, hence reducing the number of blockages. An advanced approach to commissioning was completed for this new project due to the risk associated with the installation of a complex and untested unit into a critical position in the production line.

The advanced approach to commissioning allowed for rigorous testing to be completed in a controlled environment with low risk to production capacity. Commissioning was completed in two stages with the development of a model system and full scale commissioning of the new system before installation.

Development of the model system allowed the basic concept of the system to be tested. The model was constructed out of crude materials and was tested under a range of conditions to determine the optimal operating parameters. The full scale vertical screw was designed and constructed based on the results obtained from the model system.

Commissioning was completed by operating the system under a set of conditions established to simulate normal operation. The simulation was completed by assembling the new system directly adjacent to its intended future position in the operational plant. It was wired into the system using all of the final wiring components but was not installed into the process. A feed hopper was fitted to the inlet of the screw to simulate the priority feed system and water and various other components were used to simulate the environment of the reaction chamber. The operation of the vertical screw during the advanced commissioning phase can be seen in Figure 7. The process of running the system under simulated operating conditions allowed the full commissioning of the mechanical and electrical systems. It also allowed for the partial commissioning of the operation capacity. During this process it was found that several components did not operate as expected. Changes were made to the system and were re-commissioned without any negative effect to the production capacity of the plant.

Figure 7: Advanced commissioning of vertical feed screw during plant operation. Image shows plant being fed with material via a temporary arrangement while being commissioned. (Photograph by K Lawry and used by permission of Ballance Agri-Nutrients.)

The process of commissioning prior to the installation of the vertical screw into the system was extremely successful. Full mechanical and electrical commissioning was completed as the full electrical system was used to drive the advanced commissioning process. The operational testing identified several design flaws in the vertical screw. It also reduced the uncertainty associated with the operation of this new technology.

Advanced commissioning has one significant drawback; it only reduced risk associated with the installation of the new vertical screw. The simulated process cannot represent the real process exactly. There are several factors such as continuous operation in the highly reactive environment that cannot be tested until the system is fully installed. Nonetheless the advanced approach was effective in eliminating latent defects and thereby reduced the overall risk.

Tiwai Point Ship Unloader

New Zealand Aluminium Smelters (NZAS) in Bluff installed a new ship unloader on the Tiwai wharf for the unloading of alumina and coke from incoming ships. The new unloader produced by Alesa Engineering Ltd. replaced the forty-year-old unloader that was installed when the smelter was first constructed in 1971. The new unloader has a significantly increased capacity capable of discharging at 1,000 tonnes per hour (TPH) of alumina and 600 TPH of coke compared to the old unloader that was only capable of discharging 235 TPH of alumina and 250 TPH of coke [30]. The new ship unloader was installed with the aim of reducing the time required for a ship to spend unloading. Less time spent unloading will mean a more efficient use of shipping resources and the reduction of costs associated with slow turn around (demurrage).

Installation and commissioning was completed by Alesa under the guidance of specialist project engineering company Bechtel who work onsite at the Tiwai point aluminium smelter. This process had to be completed under tight constraints as it was critical that there were no process interruptions. The smelter is a continuous process which cannot be shut down or restarted without high associated cost.

It was decided to take a parallel approach to commissioning; the old unloader must remain in a fully operational state until the new system has been thoroughly commissioned and proven capable of carrying the full operational load. Keeping the old ship unloader in an operational state significantly reduced the risk of supply interruptions, but introduced additional concerns relating to the integration of the two systems. Integration of both unloaders on the same wharf was completed by limiting the operation of the old unloader to the north half of the wharf, while the new unloader was installed and commissioned on the south half, as seen in Figure 8. Limiting the old unloader to half of the wharf increased the unloading time as the ship had to be manoeuvred around to allow access to all of the holds. This was taken as a minimal sacrifice to ensure consistent supply.

Figure 8: Parallel operation of old (left) and new (right) ship unloaders on the Tiwai wharf. (Photograph by K Lawry.)

The main modification that was required for the integration of the new unloader into the existing infrastructure was the installation of a new conveyor system to replace the southern half of the existing wharf conveyor, clearly this has cost implications. This was completed before the new unloader was installed. Both the new and old conveyor systems operated as one continuous conveyor that serviced the new and old unloaders simultaneously. The upgrade of the conveyor acted to integrate the two unloaders into the overall process. The new conveyor and ship unloader were constructed and assembled off site. The units were then transported and lifted into place. The use of pre-constructed assembled units significantly reduced installation time, therefore allowing installation to be completed in the short window between scheduled shipping movements.

Several complications emerged during the commissioning, and these extended the project duration. However the parallel approach meant that these had no impact on the overall production capacity of the smelter. The reduction in unloading capacity caused by the limitation of the old unloader to the north half of the wharf was quickly offset by the high capacity of the new unloader even when it was operating at a reduced output. Parallel commission proved to be a successful method of commissioning as there were relatively minor additional costs and the risk of process downtime was completely mitigated. The old unloader was decommissioned once the new unit was fully operational [31].

DISCUSSION

This paper makes several novel contributions. First, it provides a novel conceptual framework for the commissioning process. The model represents the decision making within the process, the broader context in which plant commissioning occurs, as well as making provision for the finer details. The novelty is creating a structured representation of the commission process, where models are otherwise sparse. Commissioning is generally an ad hoc process, and the value of this new framework is that it provides a structured theoretical foundation for this important activity.

A second contribution is the categorisation of commissioning into three main types: advanced, parallel, and direct. This exposes the variability of strategies within the commissioning process, so it becomes apparent that there is not merely one universal approach to commissioning. Achieving this adds choice to the project planning. It makes it clear that there are choices that practitioners can make, and stating these choices encourages a thoughtful consideration of the planning and resource implications thereof. This categorisation thus adds richness to the conceptual model and makes the decision points more explicit, without being prescriptive.

A third contribution is the development of a model for use by practitioners. The model captures and represents the proposed situational variables (contingency factors) involved in the process. This is valuable for informing the decision making of practitioners. The applicability of the model has been demonstrated by case studies.

A fourth contribution is the integration of commissioning into other management models. The model provides integration with the "risk management" and "project management" disciplines. This is valuable because it shows practitioners how commissioning may be approached in a more holistic manner. The commissioning model is also integrated into a wider model for the development of production plant, and thereby into "manufacturing engineering" including quality and lean manufacturing. Space does not permit full description of this integration, but the point is that the work shows that this integration is indeed feasible. The model is represented in IDEF0 notation, which is a production engineering notation, meaning that it is readily comprehendible in that context.

Overall what has been achieved is to replace the otherwise ad hoc process of commissioning with a systematic process complete with proposed decision factors and internal models of causality. There are implications for practitioners in the model, in the form of flowcharts identifying the critical success/failure factors for commissioning. Thus tentative recommendations can be made for the best commissioning approach for a given situation.

There are also implications for further research. The model is at least partly conjectural, and further research could be directed at establishing the validity of the proposed causal relationships. Another strand of research could be directed at further refinement of the model, and its extension deeper into specific cases, that is, further investigation of the situational variables.

CONCLUSIONS

Commissioning is extremely valuable to all projects but is poorly defined in the project management body of knowledge. The existing literature on commissioning is focussed on specific tasks, and holistic perspectives are lacking. This work has reconceptualised commissioning and shown that it is possible to identify three main types of commissioning (direct, parallel, and advanced) and construct a generalised conceptual framework around them. This approach to commissioning has been demonstrated by application to case studies.

The value of this work is that it presents a different and more structured way of thinking about commissioning. This allows for a more critical examination of how to complete the commission for a particular project, and ultimately the potential for a better commissioning outcome for practitioners. For theorists the benefit is that a generalised model has been developed, thus a foundation for future advancement of the subject. We have shown that the commissioning activities can be integrated into the risk management, project management, and production engineering bodies of knowledge.

ACKNOWLEDGMENTS

The authors would like to thank Ballance Agri-Nutrients and Richard Sweney at New Zealand Aluminium Smelters for providing the information required for the cases studies. These cases studies provided a valuable insight into how commissioning is completed in industry and would not have been possible without the help from these organisations.

REFERENCES

1. R. Bernhardt, "Approaches for commissioning time reduction," Industrial Robot, vol. 24, no. 1, pp. 62–71, 1997.

2. R. B. Brown, M. B. Rowe, H. Nguyen, and J. R. Spittler, "Time-constrained project delivery issues,"AACE International Transactions(PM. 09), pp. 1–7, 2001.

3. P. Gikas, "Commissioning of the gigantic anaerobic sludge digesters at the wastewater treatment plant of Athens," Environmental Technology, vol. 29, no. 2, pp. 131–139, 2008. ·

4. D. Horsley, Ed., Process Plant Commissioning, Institution of Chemical Engineers, Rugby, UK, 2nd edition, 1998.

5. IPENZ, Practice Note 09: Commissioning Capital Plant, IPENZ, Wellington, New Zealand, 2007.

6. J. Kirsilä, M. Hellström, and K. Wikström, "Integration as a project management concept: a study of the commissioning process in industrial deliveries," International Journal of Project Management, vol. 25, no. 7, pp. 714–721, 2007. ·

7. NHS, Project Management in a PCT Environment, National Primary and Care Trust Development Programme, 2004.

8. B. Peachey, R. Evitts, and G. Hill, "Project management for chemical engineers," Education for Chemical Engineers, vol. 2, no. 1, pp. 14–19, 2007. ·

9. Project Management Institute (PMI), A Guide to the Project Management Body of Knowledge (PMBOK Guide), Project Management Institute, Newtown Square, Pa, USA, 4th edition, 2008.

10. P. V. Thomas, "Best practice for process plant modifications (fertilizer plants)," Cost Engineering, vol. 45, no. 5, pp. 19–29, 2003.

11. V. S. Sohmen, "Capital project commissioning. Factors for success," in Proceedings of the 36th Annual Transactions of the American Association of Cost Engineers (AACE› 92), Orlando, Fla, USA, June-July 1992.

12. H. M. Guven and S. T. Spaeth, "Commissioning process and roles of pyers," in Proceedings of the ASHRAE Winter Meeting, la, New Orleans, La, USA, January 1994.

13. S. K. Shome, "Integration of commissioning activities in project management in power sector," inProceedings of the Project Management in the Power Sector Seminar, Ooty, India, November 1982.

14. M. G. Tribe and R. R. Johnson, "Effective capital project commissioning," in Proceedings of the 54th IEEE Pulp and Paper Industry Technical Conference (PPIC› 08), Piscataway, NJ, USA, June 2008.

15. E. E. Choat, "Implementing the commissioning process," in Proceedings of the Winter Meeting of ASHRAE Transactions, Part 1, Chicago, Ill, USA, January 1993 1993.

16. S. Doty, "Simplifying the commissioning process," Energy Engineering, vol. 104, no. 2, pp. 25–45, 2007. ·

17. E. Schepers, "Commissioning chemical process plant," in Proceedings of the 2nd National Chemical Engineering Conference, pp. 60–69, Institution of Chemical Engineers, University of Queensland, Surfers Paradise, Australia, 1974.

18. G. Shimmings, "Reflections on the causes of delays in commissioning automated materials handling projects," in Proceedings of the 3rd International Conference on Automated Materials Handling, Birmingham, UK, 1986.

19. A. Levi and M. Stonell, "Project management and commissioning of industrial boiler plant,"Institution of Mechanical Engineers, Conference Publications, pp. 55–67, 1979. ·

20. E. Cagno, F. Caron, and M. Mancini, "Risk analysis in plant commissioning: the Multilevel Hazop,"Reliability Engineering and System Safety, vol. 77, no. 3, pp. 309–323, 2002. ·

21. V. Ramnath, "How you can precommission process plants systematically," Hydrocarbon Processing, vol. 90, no. 4, pp. 119–124, 2011.

22. A. Rautenbach, "Site acceptance testing and commissioning of process control systems," Elektron, vol. 19, no. 5, pp. 40–44, 2002.

23. G. Reid, "How to achieve successful startup and commissioning for instrumentation and controls project," in Proceedings of the Advances in Instrumentation and Control Conference, vol. 47, pp. 121–124, ISA Services, Houston, Tex, USA, October 1992 1992.

24. ISO 31000, Risk Management—Principles and Guidelines, International Organization for Standardization, 2009.

25. D. J. Pons, "Strategic risk management in manufacturing," The Open Industrial and Manufacturing Engineering Journal, vol. 3, pp. 13–29, 2010.

26. J. Leitch, "Eliminating the risks to starting up your plant right the first time," Hydrocarbon Processing, vol. 85, no. 12, pp. 47–52, 2006.

27. FIPS, "Integration definition for function modeling (IDEF0)," 1993,http://www.itl.nist.gov/fipspubs/idef02.doc.

28. KBSI, "IDEF0 overview," 2000, http://www.idef.com/idef0.htm.

29. D. J. Pons, "Ventures of co-ordinated effort," International Journal of Project Organisation and Management, vol. 4, no. 3, pp. 231–255, 2012.

30. NZAS, "Unloader," in Tiwai Pointer, pp. 1–7, Newsletter of New Zealand Aluminium Smelters, 2011.

31. NZAS, "A look at our new ship unloader," in Tiwai Pointer, pp. 1–7, Newsletter of New Zealand Aluminium Smelters, 2012.

An Integrated Approach to Remanufacturing: Model of a Remanufacturing System

Ana Paula Barquet[1], Henrique Rozenfeld[1], and
Fernando A Forcellini[2]

[1]São Carlos School of Engineering, University of São Paulo (USP),
Av. Trabalhador São-carlense, 400, São Carlos, Sao Paulo, 13566-
590, Brazil

[2]Industrial Engineering Department, Federal University of Santa
Catarina (UFSC), Campus Reitor João David Ferreira Lima,
Florianópolis, Santa Catarina , 88040-970, Brazil

ABSTRACT

Remanufacturing is the process of rebuilding used products that
ensures that the quality of remanufactured products is equivalent
to that of new ones. Although the theme is gaining ground, it
is still little explored due to lack of knowledge, the difficulty of

visualizing it systemically, and implementing it effectively. Few models treat remanufacturing as a system. Most of the studies still treated remanufacturing as an isolated process, preventing it from being seen in an integrated manner. Therefore, the aim of this work is to organize the knowledge about remanufacturing, offering a vision of remanufacturing system and contributing to an integrated view about the theme. The methodology employed was a literature review, adopting the General Theory of Systems to characterize the remanufacturing system. This work consolidates and organizes the elements of this system, enabling a better understanding of remanufacturing and assisting companies in adopting the concept.

BACKGROUND

Increasing market competition, environmental concerns, changing customer requirements, and the emergence of new laws for end-of-life product management have led companies to seek new ways to maintain and expand their market share [1]. In this context, the adoption of product end-of-life strategies, which include recycling, reusing, and remanufacturing, have gained increasing importance in day-to-day business.

In reuse, according to Rose [2], the product and/or components are used immediately after their first cycle, i.e., they are second-hand goods. On the other hand, according to Thierry et al. [3], the purpose of recycling is to enable the reuse of the materials of used products and their components. In this case, the built-in energy, identity, and functionality of products and components are lost. Remanufacturing, however, preserves the shape and added value of products since the remanufactured product should be used for the same purpose it had during its original life cycle [4]. In remanufacturing, the used product returns to the production line, where it is disassembled, cleaned, reconditioned, inspected, and reassembled to ensure that the remanufactured product has the same quality as a new one. Additionally, less effort and resources are required for the recovery of products and their components compared with other product end-of-life strategies [5].

Among these strategies, remanufacturing is one of the preferable alternatives [6] since the remanufacturing process preserves part of the raw materials and value added to the product during its fabrication, allowing companies to increase their productivity and profitability [7]. Hence, in view of its environmental and economic benefits, remanufacturing is gaining significant ground in the global scenario.

However, it is difficult to achieve an integrated and systematic vision of all the issues involved in remanufacturing. Remanufacturing is a complex business due to the high degree of uncertainty in the production process, mainly caused by two factors: the quantity and the quality of returned products. Clearly, the lack of an integrated perspective of remanufacturing limits the possibilities for companies to evaluate and decide about whether or not to offer remanufactured products [8].

The potential of remanufacturing is underexploited in Brazil. There are still only few Brazilian companies showing awareness of environmental issues and commitment to the fate of the used products they manufacture, although this situation is expected to change soon pursuant to the newly enacted National Policy on Solid Waste. Acting to change this situation, these companies can increase the degree of competitiveness *vis-à-vis* foreign companies, gain new customers, and survive in the market, as well as contribute toward more sustainable production and consumption.

Companies find it difficult to implement and consolidate remanufacturing for several reasons, including a lack of knowledge about the theme, a lack of consideration about the strategic issues of remanufacturing [8], and the scantiness of studies indicating how to implement it [6]. Thus, it is clear that this theme, which is a new one especially for Brazilian companies, is in its exploratory phase, which explains the importance of a structured review of the literature about remanufacturing.

In this paper, remanufacturing is treated as a system. A system is considered a set of interdependent elements that interact to achieve an objective and perform a given function. It is the elements and their relationships to each other that determine how the system works, forming a unitary, organized, and complex whole [9]. The majority of authors discuss isolated elements of remanufacturing, making it difficult to gain an integrated view of them, which are treated separately and in different contexts. It is believed that the conceptuation of a model for the remanufacturing system can help companies understand and implement remanufacturing. In agreement with Östlin [10], characterizing the remanufacturing system contributes substantially toward understanding the problems and difficulties involved in remanufacturing.

Thus, the objective of this work is to organize the body of knowledge about remanufacturing by means of a model that offers a vision of the remanufacturing system, the elements in this system, and how they interact with each other, contributing toward an integrated view of the theme.

This paper is divided into introduction and context. The next section presents a literature review about remanufacturing, with the findings organized so as to draw up a model of the remanufacturing system. The elements of this system are characterized, and their difficulties and practices described, as well as their interdependencies and interconnections.

METHOD

This work was developed based on General System Theory [11] and by means of a cross-analysis of the elements and characteristics of remanufacturing found in the literature review. Since we intend to present these elements within a systemic vision, we use the General System Theory, the objective of which is to study the elements that make up a system as well as the interactions between them [11]. Studying each element separately does not lead to an exact conclusion of the system in which these elements are inserted for their interactions are fundamental to understanding the system as a whole. As we found in the literature, this is the case of papers that deal with remanufacturing, which in the most part discuss its elements separately. General System Theory arose from the need to understand the problems of today's complex world; however, analyzing them separately and dealing with them piecemeal to fit theoretical problems and problems resulting from modern technology does not suffice. A system, or 'organized complexity', can be defined as a set of elements governed by 'strong interactions' [11]. Uhlmann [12], based on Bertalanffy, sees a system as a set of elements interrelated to each other and to the environment. The next section presents our literature review about remanufacturing, in which concepts are organized to define and characterize the remanufacturing system and the elements that make up this system.

Remanufacturing System

An organizational system can be considered a set of dynamic and interdependent parts and functions with shared objectives. These systems are open and may belong to larger systems and contain smaller ones. They present specific objectives and complex structures. Since a system is larger than the sum of its parts, the investigation of any part of a system should involve it as a whole.

The remanufacturing system proposed here is inspired by Östlin [10]. According to this author, this system begins with the collection of the used product or parts, also named as core, followed by its remanufacturing and delivery of the remanufactured product to the client. Thus, the remanufacturing system comprises internal processes, such as the remanufacturing operation, and external processes involving the collection of cores and delivery of the remanufactured product.

Seeking to complement Östlin's proposal [10], this paper proposes the following elements and sub-elements for the remanufacturing system:

- Element 1: Design for remanufacturing
- Element 2: Reverse supply chain (RSC)
  ○ Sub-element 2.1: Acquisition/relationship with the core supplier
  ○ Sub-element 2.2: Reverse logistics (RL)
- Element 3: Information flow in the remanufacturing system

- Element 4: Employees' knowledge and skills in remanufacturing
- Element 5: Remanufacturing operation
- Element 6: Commercialization of the remanufactured product

Figure 1 illustrates the proposed remanufacturing system. The first element is the design for remanufacturing, which is part of the product development process and is responsible for the product's design, with a view to its end of life and how to facilitate its remanufacturing (e.g., facilitate disassembly). This element provides information for the reverse supply chain, particularly for element 5 (remanufacturing operation; e.g., disassembly sequence). The reverse supply chain, in turn, is composed of two sub-elements: acquisition/relationship with the core supplier and reverse logistics. The last element is the commercialization of the remanufactured product. Among these elements, there are both information and material flows. The main actor of this system is the end client, who becomes the supplier at the end of life of the product.

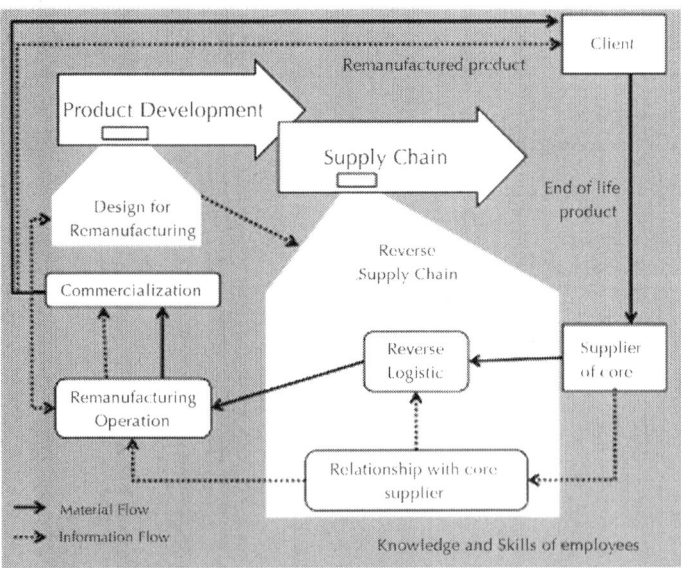

Figure 1: Remanufacturing system model (adapted from [13]).

Given the importance and influence of the information flow throughout the remanufacturing system, it is also considered an element that permeates and interconnects all the others. The flow of materials was not mentioned as an element because it is mainly part of the sub-elements of the reverse supply chain. It should also be kept in mind that a crucial factor for the feasibility of the system is the employees' knowledge and skills in remanufacturing (element 4), which are required in all the processes. Each of the aforementioned elements is characterized in the following items, describing their characteristics, difficulties, and practices.

Element 1: Design for Remanufacturing

Characteristics

Approximately 80% of the environmental impacts of products are determined during their development, more specifically in the concept phase, which emphasizes the responsibility of product development teams to address issues related to the service life of products [14]. This underlines the importance of designing the product considering the most suitable end-of-life strategies for their reuse after their use, such as remanufacturing [15].

A variety of pressures challenge companies to alter their product development paradigms. The sanction of legislation on manufacturer responsibility, allied to increasing global competition and the potential for the recovery of used products to make use of their residual value, encourages companies to design products with greater durability and facility to reuse them at their end of life [16].

For Andrue (in [16]), remanufacturable products have the following characteristics:

- The product contains a component or part that allows for its reuse.
- There is availability in the supply of such components or parts.
- The product and/or its parts can be disassembled and reused

according to the original specifications.

- The product and/or its parts has high added value in relation to its market value and its original cost.
- The product and the process are stable.
- Examples of the product's characteristics that affect remanufacturing, identified by Ijomah et al. [16], are listed below:
- Technological changes: At the time the product is to be remanufactured, its technology may have become obsolete. If the product cannot be updated during remanufacturing, its reuse is unnecessary.
- Business model of services: Development of business model that allows for a combination of products and services, e.g., a product service system.
- Environmental legislation: This type of legislation may require companies to reuse the product at its end of life and make it more expensive to discard, for instance.

During product conception, it is important to take into account strategies for updating the product due to rapid changes in technology as well as clients' needs. In this case, too, design for remanufacturing can give companies a significant market advantage [15]. Integrating the design for remanufacturing in the product development process optimizes the achievement of the benefits of remanufacturing by companies (reduction of energy, materials and wastes, among others). This is the case of companies such as Xerox Corporation (Norwalk, USA), which recognizes this factor as an opportunity to obtain competitive advantages [15].

Difficulties and Practices

Many of the challenges related to remanufacturing are consequences of how the products were designed [16,17]. In order to support the company's product development process, several ecodesign methods and tools have been developed, which consider remanufacturing and product end-of-life issues [18].

Interdependence and Interaction among the Elements

A limiting factor for putting design for remanufacturing into practice is the low level of knowledge of product designers concerning about end-of-life strategies, such as a remanufacturing. This is due to the fact that a product's conception is usually centered on its functionality and costs, in detriment to environmental issues [16]. Thus, this indicates a relationship between the elements' design for remanufacturing and employees' knowledge and skills.

It should also be noted that to perform design for remanufacturing, product designers require specific expertise to develop the product with a view to its easy future remanufacture, in other words, to facilitate the efficient execution of the remanufacturing operation.

Element 2: Reverse Supply Chain

Characteristics

According to Guide and Van Wassenhove [19], RSC is a set of stages needed to collect the core, followed by the application of the desired end-of-life strategy (remanufacturing, reuse, recycling or disposal) [1]. Driven by cost reductions through product reuse and by the customer's heightened perception of value, many supply chains have increased their involvement in activities that go beyond the product's service life, extending its life cycle [20].

The RSC begins with the collection of products from clients and/or companies in different links of the supply chain, and collection sources tend to be geographically dispersed. The next phase involves inspection or tests performed at the collection site, at a receiving center, or at the site where the product will be reused. At this point, a decision is made about the destination of the collected product, which presents various possibilities of reuse, such as remanufacturing [21].

An example of how a reverse supply chain works is described by Guide and Van Wassenhove [22]. To reuse used mobile phones, it is necessary first to gain access to a sufficient number of telephones of suitable quality and at the right price (acquisition of the used product). The telephones must be transported and stored (reverse logistics); after which, an end-of-life strategy is selected for the product (including remanufacturing). After they have been remanufactured (remanufacturing operation), product commercialization strategies are devised. Some recommendations to achieve efficiency in the reverse supply chain are the following [23]:

- Structure a team specialized in contacts with the used product supplier, with a view to standardizing the processes involved in the collection and increasing these products' chances for reuse;
- Make a forecast of the time of return of the product, which may be based on its sales and its service life;
- Align the reverse supply chain with the direct supply chain and achieve the effectiveness of the activities of a closed-loop supply chain.

Difficulties and Practices

The phases of the RSC are treated as a series of independent stages that are dealt with separately, without considering their integrated nature. Moreover, business and academia are doing very little about the strategic issues of the RSC [1]. The majority of studies are centered on technical and operational issues since the focus on technical activities is attractive for initial investigations [22].

Because RSCs are not yet part of the core competencies of companies, it is difficult to organize and align their stages, obtain the necessary resources, and catch the attention of top management. The direct supply chain requires similar requisites, but the issues and the context of reverse supply chains are still little understood, more complex, and receive scanty attention [1]. The reverse supply chain presents some characteristics that make managing and planning its stages and activities complex. For example, when a company collects cores, its supplier is usually the end client, which makes it difficult to gain access to a sufficient number of cores at the moment of return. Another complication is the quality of these cores, which requires efficient inspection [24]. In addition, there is the need to disassemble the collected products, to set up a reverse logistics network, and to deal with the high variability in processing times [25].

Sub-elements of the RSC

The sub-elements of the RSC considered in this paper, which were adapted from the works of Guide and Van Wassenhove [22], and Blackburn et al. [26], are the following:

- Acquisition/relationship with the used product supplier: contact with the supplier for the acquisition of cores by the remanufacturer;
- Reverse logistics: the activities of transport, storage, and distribution of the products that will be reused.

Acquisition/Relationship with the used Product Supplier

- Characteristics. Sundin et al. [27] concluded that the remanufacturer's relationship with his supplier of cores is an extremely important aspect for the effectiveness of the business, which is consistent with the findings of Östlin et al. [28].
- Difficulties and practices. If the collected product is destined for remanufacturing, it is important to emphasize that before the product is selected for return by the reverse flow, its remanufacturability should be 'pre-assessed' to avoid it being transported without it serving for reuse, which would incur additional costs. This trade-off between potential reuse and additional costs is difficult and requires employees with certain skills and experience [29]. Employees that perform

this pre-assessment should be part of the team responsible for collecting cores because if the product's remanufacturability is evaluated at the remanufacturing company and this product is deemed unfit for reuse, its transportation to the company will have been unnecessary.

The large number of suppliers of small quantities of cores and the diversity of their conditions makes it difficult for the companies that receive them for remanufacturing to control their quality, as does the lack of closeness in the relationship between the remanufacturer and his supplier. Moreover, when selecting their suppliers, many remanufacturers choose the ones that offer the lowest prices[6]. This may lead to the purchase of cores in poor conditions to be remanufactured.

In some cases, the suppliers of cores are the clients that discard the product due to its end-of-service life or for other reasons, such as the launch of a more modern product. In this case, there are some problems concerning the lack of motivation to get these customers to return used products to remanufacturing companies [30]. Below are some situations that may influence this lack of motivation:

- The remanufacturer's lack of contact with the client [28];
- The client's lack of knowledge and confidence about remanufacturing [31], e.g., the economic and environmental advantages of remanufacturing.

Thus, companies that want to succeed should think about remanufacturing strategies that encourage the client to make this return and that bring him close to the remanufacturing company [28].

Reverse Logistics

1. Characteristics. According to the Council of Logistics Management [32], reverse logistics is the process of planning, implementing, and effectively controlling the flow of components, materials undergoing processing, end product, and related information from the consumption point to the source point [23].

2.   Difficulties and practices. Reverse logistics is one of the great challenges of the remanufacturing system due to the difficulty in predicting product volumes, return times, and quality conditions, which makes planning of the remanufacturing operation difficult [24]. In recent years, RL has received more attention from managers due to its strategic implications [33].

Some of the issues that make it difficult for companies to implement RL are the lack of a system that integrates the activities of direct and reverse logistics [34], the difficulty of measuring the impact and of controlling the return of products and materials, and the fact that reverse flow is considered a cost for companies and therefore is given little or no priority as a business strategy [32]. The costs of reverse logistics activities can rely on different actors. The company that will remanufacture can be the one to handle with this cost by acquiring cores directly from clients. In a different scenario, these cores can be stored in deposits by an intermediate actor that will sell them to the one responsible for performing remanufacturing. In both cases, there is a great probability that the costs of reverse logistics activities are added to the remanufactured product final costs.

Hence, most companies are uninterested in implementing RL. In addition, there is a lack of studies by companies to assess the effects of the practice of RL on the success of organizations, the relationship between the actors involved in the activities of reverse logistics is poorly structured, there is little closeness between the plants and the suppliers of cores, and the cost of shipping is higher due to the lower volumes transported (Ballou 2006 in [35]).

Unlike distribution in direct logistics, which is designed to transport large volumes of the same product from the manufacturer to a few local clients, in reverse logistics, the product mix may vary considerably and the volume may be very low. This can make economic transportation difficult to achieve [29].

Another aspect to consider is the fact that, contrary to the situation in direct logistics, cores are not packaged and are therefore unprotected and susceptible to damage, which limits their recovering. Therefore, a packaging system is necessary to protect the product's residual value [29]. Pires [35] points out some characteristics specific to RL, which affect its efficiency:

- A convergent network structure, i.e., products from numerous sources and with few destinations;
- Geographically dispersed sources, usually not homogeneous in quantity, availability, and quality of product and/or parts with each other and over time, making planning activities difficult;
- Difficulty in achieving an economy of scale due to the small quantity of products collected from each source;
- Higher tendency for products to stay longer in reverse channels, resulting in higher inventory, transport, and storage costs, as well as reduced income due to the possibility of product obsolescence and degradation;
- Entrance of products into the flow that should not enter (e.g., products that cannot be reused), generating unnecessary costs;
- When the client is the core supplier, RL depends on his willingness to cooperate and reinsert the post-consumer material into the reverse logistics flow.

For Lacerda [36], the main factors that influence the efficiency of reverse logistics activities are good input and output controls, mapped and formalized processes, short cycle times, accurate information systems, planned logistics network, and collaborative relationships between clients and suppliers.

Element 3: Information Flow in the Remanufacturing System

Characteristics

The main role of the information flow in the remanufacturing system is to deal with uncertainties concerning the return of products. An efficient information flow is an important tool to reduce these uncertainties and to help establish an effective system.

Due to the characteristics of the remanufacturing system, the information required for planning the system becomes accessibly very late, hampering its operation. If the remanufacturer does not reduce the lead time of information, the coordination of the remanufacturing operations and reverse logistics is impaired, thereby increasing the costs of the system [6]. Information about the product is important both in planning the reuse activities and in avoiding the transportation and reprocessing of products that have no potential to be recovered.

Difficulties and Practices

To reduce these uncertainties and their consequences on the remanufacturing system, the remanufacturer must manage the following information [6]:

- Which products should be returned to the remanufacturer?
  When will these products arrive?
  Where are these products located?
- How many of these products can be remanufactured?
- According to Thierry et al. [3], information related with product return management can be classified in four categories:
- Information about the composition of the product: types of material, their quantities, value, potential of harmfulness to nature, and how the different types of materials are combined.
- Information about the magnitude and uncertainty of the return flow: according to the type of commercialization chosen for the product, e.g., traditional sale, leasing, rental.
- Information about the market for remanufactured products: the perceived difference in the quality and cost of remanufactured and new products affects the acceptance of these products.
- Information about how product returns are currently done includes an analysis of the organizations involved, the obstacles, and the quantity of product that is remanufactured (for each returned product), the costs, and the overall environmental impact of the remanufacturing system.

Most companies experience difficulties in obtaining accurate information about aspects of product return management in their supply chains. However, companies can obtain such information by collaborating with their suppliers and others in the chain. To obtain this information, it is essential to deal with issues involving the characteristics of the product, the supply of cores, and the demand for remanufactured products, and to balance the supply and demand. Obtaining and processing this information require the development of an adequate information system [3].

Element 4: Employees' Knowledge and Skills in Remanufacturing

Characteristics

From the moment of contact with the supplier, acquisition, collection of cores (transport, storage), and the phases of the remanufacturing operation (inspection, disassembly, cleaning, etc.), to the sale of the remanufactured product, the employees should be perfectly familiar with the remanufacturing system in order to deal correctly with each stage of the system.

Difficulties and Practices

According to Jacobsson [6], the success of the implementation of the remanufacturing operation often does not require a more highly qualified work force than the one in the manufacturing operation, but the qualifications required in each of these operations are different. In the remanufacturing operation, the employees should be trained and qualified to deal with variability and uncertainties, especially insofar as quality and quantity of cores are concerned.

Uncertainties lead to situations in which resources will sometimes be scarce and at other times abundant, thus requiring people with a good vision of the overall operation and with the necessary flexibility to deal with the different stages of the operation [37]. For the aforementioned authors, both qualified and non-qualified employees are necessary since non-qualified people are usually better able to think outside the box and contribute with new ideas.

Training for the remanufacturing operation should preferentially be given by the remanufacturing company itself which can train its employees according to the specificities of its operations and the characteristics and complexity of the product. For example, employees responsible for disassembling products should take them apart without damaging them. Cleaning, handling the product, and assembling it require less specialized qualifications, less precision, experience, and skills than inspection and testing [6].

Element 5: the Remanufacturing Operation

Characteristics

The remanufacturing operation begins with the arrival of the core at the remanufacturer's facilities, where it will go through several stages that include its complete disassembly, cleaning of its parts, inspection, reconditioning of the parts that will be reused, replacement of non-remanufacturable components, and assembly, resulting in a remanufactured product. This product is then tested to ensure that its quality is equivalent to that of a new product [16,31,38]. The order of these activities may differ according to the characteristics and type of the product [16]. It should be noted that, in this paper, the steps of reconditioning and replacement of used parts for new ones, with or without possible product updates, will be called reprocessing.

Next, a brief description of the stages of remanufacturing operation is given, according to Steinhilper[39]:

- Product disassembly: The purpose of this stage is the total disassembly of the product. This is one of the most time-consuming activities due to its degree of complexity.

- Cleaning of the components: Each of the components is cleaned with a different cleaning product, according to the material composing it. Four process variants can be cited that contribute to this cleaning: chemical effects (e.g., detergents), influence of temperature (e.g., heat), mechanical action (e.g., removal by high-pressure water jetting), and time (e.g., duration of the process).
- Inspection and storage of components: This stage proposes an identification to classify the components as well as an inspection to determine which should be replaced or allocated for other purposes, such as cannibalization, repair, reconditioning, or recycling. Storage refers to the site where the material will be deposited for subsequent assembly of the products.
- Reconditioning and replacement of components and parts (reprocessing): Components and parts are recovered. Therefore, some of them are replaced with new ones because they do not satisfy the minimal requirements to ensure the quality of the remanufactured product.
- Product reassembly: This consists of the assembly of the remanufactured product. A final test will ensure that the remanufactured product performs similarly to a new one, with the same characteristics, functionalities, and quality.

Steinhilper [39] states that the final test should not be considered a step but a certification that the product will have the same characteristics as a new one. The stages of the remanufacturing operation may follow a different order, depending on the type of product to be remanufactured.

Difficulties and Practices

Remanufacturing operations require lower investments than manufacturing operations since the number of new parts produced is smaller, and a large part of the efforts and resources required has already been invested by the original manufacturer (Lund and Skeels 1983 in [4]).

Element 6: Commercialization of the Remanufactured Product

Characteristics

This element encompasses sales, distribution, and relationship with the client who bought the remanufactured product. Additionally, it explores potential market segments and strategies to increase the attractiveness of remanufactured products.

Remanufactured products can be commercialized in different ways. Some companies may choose to sell them, while others will turn to leasing, using them as replacement products for warranties or selling their functionality to the client. The clients' preferences, the nature of the product, and its technical maturity are the main factors that influence the decision about which channel to use.

Atasu et al. [40] conducted a study involving a marketing approach to remanufacturing, related with the market demand for remanufactured products. These authors treat remanufacturing as a strategic marketing tool that strongly enhances the company's competitive advantage, which differs from the idea of remanufacturing as cost savings or an obligation with legal implications. The authors concluded that the main factors that influence the decision to remanufacture are competition, market growth, and increase in the 'green' range of the market.

An important decision about the commercialization of remanufactured products is whether or not they will be sold through the same distribution channels as new products. Because remanufactured products cost less than new products, they may cannibalize the sales of new ones. Therefore, many companies do not offer remanufactured products together with new ones. This is the case of Dell, Inc. (Round Rock, USA), which has a separate site for the sale of remanufactured products (http://www.delloutlet.com *webcite*). On the other hand, selling remanufactured products together with new ones enables companies to better segment their market and to sell also to clients who cannot afford new products [41].

Difficulties and Practices

Few studies explain how to place remanufactured products on the market [6]. In this regard, remanufacturers face major challenges because even though there is an increasing demand for environmentally attractive products, the remanufactured product contains parts, components, or materials that have been used previously. Therefore, marketing strategies must be adapted to market this product, especially with respect to issues such as below [6]:

- Market segment: Lower prices for products with the same performance as new ones expand the market range by making them accessible to clients who cannot afford new original products.
- Buying behavior: The remanufactured product offers the same functionality as a new product, at a lower price, but may not offer the client the same shopping experience.
- Client profile: Issues involving 'fashion' and 'currentness of the fashion' also affect remanufacturing. Clients may prefer the latest version of the product, regardless of the quality and cost of the remanufactured alternative. Some clients demand novelty, and the selection and purchase of new products is a lifestyle.
- Perception about the product: Even if the price of remanufactured products is lower, some clients are not interested in them because of a perceived risk.
- Warranty: Remanufactured products should come with a guarantee that they meet the client's needs just like a new product would.
- Complementary services, e.g., warranties and maintenance.

From the standpoint of how the remanufactured product is commercialized, it has been demonstrated that the product-service system (PSS), through leasing or the offer its functionality, appears to be a promising approach [27]. Therefore, from the marketing point of view, products whose function is prioritized by the client are the most suitable ones for remanufacturing [6].

PSS can be defined as the result of a strategic innovation in the business model of companies, in which the focus shifts from the design and sale of physical products to the offer of a system of products and services that, together, can meet customer expectations. Moreover, the value is functionality rather than the physical properties of individual products [42].

The PSS is based on a fundamental change in the relationship between the manufacturers and consumers of a product and/ or service. Instead of focusing on the traditional form of sale, consumption, and disposal of the product, the PSS focuses on the delivery of a function to the client, which means a combination of products and services that, together, meet the client's needs [43].

Another important point is the need to gain the client's confidence in the remanufactured product. In a case study with a manufacturer of copier machines, which remanufactures its products, it was observed that the sales department faces challenges to persuade customers of the equivalent quality of remanufactured and new copiers [3]. Thus, ensuring the performance, reliability, and quality of remanufactured products is essential to creating and sustaining the demand for them.

RESULTS AND DISCUSSION

Analyzing the literature and characterizing the elements of the system, one sees a considerable interdependence and interaction among the elements. Firstly, with regard to the RSC, several points are relevant.

The degree of structuring of a RSC can be defined based on the existence of structured organizational practices of core returns, the relationship and information exchanged among companies that belong to the reverse chain, and the level of resources these companies make available, e.g., employees' skills in remanufacturing [6,21,32,44]. The proper management of reverse chains also serves as an excellent source of information about clients' expectations and habits, contributing for the company to provide differentiated services and to increase the value perceived by its clients [21].

A relationship with the core supplier can augment information about the product's remanufacturing conditions, since the remanufacturer will have knowledge about the performance of the product during its service life. This knowledge is useful for the remanufacturing operation as well as for the improvement of the conception of the product as a whole and for remanufacturing (design for remanufacturing).

Still with regard to the RSC, the activities of reverse logistics require skills and information to correctly carry out the transportation, storage, and warehousing of cores. Information about the volume, condition, and time of return of products is also essential for planning the remanufacturing operation.

It is essential for the employees involved in the commercialization of the remanufactured product to be properly trained, to inform the client about the benefits of the product, about what remanufacturing is and how it works, and to dispel doubts regarding the quality of the product [41].

It is also clear that an efficient information flow is necessary to underpin the relationship of the core supplier with the remanufacturer, since information about the availability and quality of the core is essential for the remanufacturing system to work properly, particularly for the planning of remanufacturing operations. Information flow between the remanufacturer and the designers of the product is also essential in case of doubts emerging during the remanufacturing operation about how the product was designed, or even for suggestions to be made to the designers, aimed at improving the design for remanufacturing.

It is also clear that the skills, knowledge, and experience of employees involved in the stages and activities of the remanufacturing system are important. The literature places particular emphasis on the importance of employees qualified for the steps of the remanufacturing operation, since this is when the used product undergoes the transformations needed to turn it into a remanufactured product with the same quality as a new one [6,37,45].

In addition, the type of product and its complexity, which are issues of design for remanufacturing, influence the sequence and difficulty of the stages of the remanufacturing operation [38]. Jacobsson[6] also mentions the importance of the qualifications of the work force for the design for remanufacturing and to deal with the financial and legal aspects of remanufacturing.

CONCLUSIONS

This work involved the organization of the body of knowledge about remanufacturing by conceptualizing the remanufacturing system and its elements, contributing toward an integrated vision and expanding the theoretical knowledge about the theme. The elements of this system were consolidated and organized, enabling a better understanding of remanufacturing and facilitating the work of future studies as well as of companies that are restructuring their remanufacturing operations or intend to start them.

This paper clearly shows the interactions and interdependence among the elements of the remanufacturing system. Dividing the elements in this paper was not intended to omit or conceal these interactions, but instead, to make it easier to understand and organize the remanufacturing system.

In terms of its academic contribution, this paper aims to promote knowledge about remanufacturing and the development of studies on the theme, particularly in Brazil, since there are still very few companies that remanufacture and few studies about remanufacturing in the country. It is also hoped that companies will feel encouraged to implement remanufacturing, since this research describes characteristics and provides a better overall understanding about the remanufacturing system.

AUTHORS' CONTRIBUTIONS

APB has done the literature research on the main publications concerning the remanufacturing topic. HR and FAF supported on the development of this article. All the authors read and approved the final manuscript.

ACKNOWLEDGMENTS

The authors would like to extend sincere thanks to Conselho Nacional de Desenvolvimento Científico e Tecnológico (CNPq) and Coordenação de Aperfeiçoamento de Pessoal de Nível Superior (CAPES) for supporting this research topic.

REFERENCES

1.   Guide VDR Jr, Harrison TP, Van Wassenhove LN: The challenge of closed-loop supply chains. *Interfaces* 2003, 33(6):3-6.

2.   Rose C: *Design for environment: a method for formulating product end-of-life strategies*. Stanford University, Dissertation; 2000.

3.   Thierry M, Salomon M, Nunen JAEE, Van Wassenhove LN: Strategic issues in product recovery management. *Calif. Manage. Rev.* 1995, 37(2):114-135.

4.   Amezquita T, Hammond R, Salazar M, Bras B: Characterizing the remanufacturability of engineering systems. In *Proceedings of ASME Advances in Design Automation Conference*. Boston, Massachusetts; 1995.

5.   Lindahl M, Sundin E, Östlin J: Environmental issues within the remanufacturing industry. In *Proceedings of LCE: 13th CIRP International Conference on Life Cycle Engineering, Katholieke Universiteit Leuven*. Belgium; 2006.

6.   Jacobsson N: *Emerging product strategies: selling services*

of remanufactured products. Lund University, Dissertation; 2000.

7. Giuntini R, Gaudette K: Remanufacturing: the next great opportunity for boosting US productivity.*Bus Horiz* 2003, 46(6):41-48.

8. Ferguson ME, Toktay LB: *The effect of competition on recovery strategies*. INSEAD, Fontainebleau; 2004.

9. Johnson RA, Kast FE, Rosenweig JE: *The Theory and Management of Systems*. McGraw-Hill, New York; 1963.

10. Östlin J: *On remanufacturing systems: analyzing and managing material flows and remanufacturing processes*. Linkoping University, Dissertation; 2008.

11. Bertalanffy VL: *Teoria Geral Dos Sistemas*. Vozes, Petrópolis; 1975.

12. Uhlmann GW: *Teoria Geral Dos Sistemas: Do Atomismo Ao Sistemismo*. Instituto Siegen, São Paulo; 2002.

13. Barquet APB, Rozenfeld H, Forcellini FA: Remanufacturing System: characterizing the reverse supply chain. In *Adaptation and Value Creating Collaborative Networks*. Edited by Camarinha-Matos LM, Pereira-Klen A, Afsarmanesh H. Springer, Heidelberg; 2011:556-563.

14. Park M: Sustainable consumption in the consumer electronics sector: design solutions and strategies to minimize product obsolescence. In *Proceeding of the 6th Asia Pacific Roundtable for Sustainable Consumption and Production*. Melbourne; 2005.

15. Nasr N, Thurston M: Remanufacturing: a key enabler to sustainable product systems.In *Proceedings of LCE. 13th CIRP International Conference on Life Cycle Engineering, Katholieke Universiteit Leuven*. Belgium; 2006.

16. Ijomah WL, McMahon CA, Hammond GP, Newman ST: Development of design for remanufacturing guidelines to support sustainable manufacturing.*Robot Comput Integrated Manuf* 2007, 23(6):712-719.

17. Hatcher GD, Ijomah WL, Windmill JFC: Design for remanufacture: a literature review and future research needs.*J Clean Prod* 2011, 19:17-18.

18. Pigosso DCA, Zanette ET, Guelere Filho A, Ometto A, Rozenfeld H: Ecodesign methods focused on remanufacturing.*J Clean Prod* 2010, 18:21-31.

19. Guide VDR Jr, Van Wassenhove LN: The reverse supply chain. *Harv Bus Rev* 2002, 80(2):25-26.

20. Corbett C, Savaskan C: Contracting and coordination in closed-loop supply chains. In*Quantitative Models for Closed Loop Supply Chain Management*. Edited by Dekker R, Fleischmann M, Inderfurth K, Wassenhove LN. Springer, New York; 2002:1-23.

21. Kopicki R, Berg MJ, Legg L: *Reuse and recycling-reverse logistics opportunities*. Oak Brook, United States; 1993.

22. Guide VDR Jr, Van Wassenhove LN: The evolution of closed-loop supply chain research.*Oper Res* 2009, 57(1):10-18.

23. Fioravanti RD, Carvalho MFH: *Aplicações de modelos de cadeia reversa em uma operação de serviços: estudo de caso no setor de serviços de impressão. XI SIMPOI - Symposium on Production, Logistics and International Operations Management, FGV.* São Paulo, Brazil; 2008.

24. Guide VDR Jr: Production planning and control for remanufacturing: industry practice and research needs.*J Oper Manag* 2000, 18:467-483.

25. Guide VDR Jr, Jayaraman V, Linton JD: Building contingency planning for closed-loop supply chains with product recovery.*J Oper Manag* 2002, 21(3):259-279.

26. Blackburn JD, Guide VDR Jr, Souza GC, Van Wassenhove LN: Reverse supply chains for commercial returns.*Calif Manage Rev* 2004, 46(2):6-22.

27. Sundin E, Ostlin J, Rönnbäck AÖ, Lindahl M, Sandström GÖ: Remanufacturing of products used in product service system offerings. In *Proceeding of the 41st CIRP conference on manufacturing systems*. Tokyo, Japan; 2008.

28. Östlin J, Sundin E, Björkman M: Importance of closed-loop supply chain relationships for product remanufacturing.*Int J Prod Econ* 2008, 115:336-348.

29. Ferrer G, Whybark DC: From garbage to goods: successful remanufacturing systems and skills.*Business Horizons* 2000, 43(6):55-64.

30. King AM, Burguess SC: The development of a remanufacturing platform design: a strategic response to the directive on waste electrical and electronic equipment.*Proc IMechE Part B: J Eng Manufacture* 2005, 219:623-631.

31. Seitz MS: A critical assessment of motives for product recovery: the case of engine remanufacturing.*J Clean Prod* 2006, 15:1147-1157.

32. Rogers DS, Tibben-Lembke RS: *Going Backwards: Reverse Logistics Practices and Trends*. Reverse Logistics Executive Council, Reno, Nevada; 1998.

33. Daugherty PJ, Autry CW, Ellinger AE: Reverse logistics: the relationship between resource commitment and program performance.*J Bus Logist* 2001, 22(1):107-123.

34. Daher CE, Silva EPS, Fonseca AP: Logística reversa: oportunidade para redução de custos através do gerenciamento da cadeia integrada de valor.*Brazilian Business Review* 2006, 3(1):58-73.

35. Pires N: *Modelo para a logística reversa dos bens de pós-consumo em um ambiente de cadeia de suprimentos*. Universidade de Santa Catarina, Thesis; 2007.

36. Lacerda L: *Logística Reversa, uma Visão sobre os Conceitos Básicos e as Práticas Operacionais*. Center of Logistics Research, Rio de Janeiro; 2004. |

37. Hermansson H, Sundin E: Managing the remanufacturing organization for an optimal product life cycle. In *Proceedings of the fourth international symposium on environmentally conscious design and inverse manufacturing, Tokyo, 12–14 December 2005*. Edited by Yamamoto R. IEEE, New York; 2005:143-156.

38. Sundin E: *Product and process design for successsful remanufacturing.* Linkoping University, Thesis; 2004.

39. Steinhilper R: Remanufacturing: the ultimate form of recycling. The Remanufacturing Institute.1998.http://www.reman.org/Publications_main.htm *webcite.* Accessed May 23, 2012

40. Atasu A, Sarvary M, Van Wassenhove LN: Remanufacturing as a marketing strategy.*Manag Sci* 2008, 54(10):1731-1746.

41. Ovchinnikov A: Revenue and cost management for remanufactured products.*Prod Oper Manag* 2011, 20(6):1-17.

42. United Nations Environment Programme (UNEP): *Product-service Systems and Sustainability. Opportunities for Sustainable Solutions.* UNEP, Paris; 2002.

43. Goedkoop MJ, Van Halen CJG, Riele HRMT, Rommens PJM: *Product Service Systems: Ecological and Economic Basics.* VROM and Economic Affairs, Netherlands; 1999.

44. Leite PR: *Logística Reversa: Meio Ambiente e Competitividade.* Prentice Hall, São Paulo; 2003.

45. Ferrer G: Yield information and supplier responsiveness in remanufacturing operations.*Eur J Oper Res* 2003, 149:540-556

Risk Determinants of Small and Medium-sized Manufacturing Enterprises (SMEs) - an Exploratory Study in New Zealand

[1]Department of Industrial and Production Engineering, Shahjalal University of Science and Technology, Sylhet, 3114, Bangladesh[2]

Aman Islam[1] and, Des Tedford[2]

[2]Department of Mechanical Engineering, The University of Auckland, Auckland, 1142, New Zealand

ABSTRACT

The smooth running of small and medium-sized manufacturing enterprises (SMEs) presents a significant challenge irrespective of the technological and human resources they may have at their disposal. SMEs continuously encounter daily internal and external undesirable events and unwanted setbacks to their operations that detract from their business performance. These are referred to as 'disturbances' in our research study. Among the disturbances, some are likely to create risks to the enterprises in terms of loss of production, manufacturing capability, human resource, market share, and, of course, economic losses. These are finally referred to as 'risk determinant' on the basis of their correlation with some risk indicators, which are linked to operational, occupational, and economic risks. To deal with these risk determinants effectively, SMEs need a systematic method of approach to identify and treat their potential effects along with an appropriate set of tools. However, initially, a strategic approach is required to identify typical risk determinants and their linkage with potential business risks. In this connection, we conducted this study to explore the answer to the research question: what are the typical risk determinants encountered by SMEs? We carried out an empirical investigation with a multi-method research approach (a combination of a questionnaire-based mail survey involving 212 SMEs and five in-depth case studies) in New Zealand. This paper presents a set of typical internal and external risk determinants, which need special attention to be dealt with to minimize operational risks of an SME.

BACKGROUND

In the dynamic and highly competitive business environment, manufacturing industries are under tremendous pressure due to the free market economy, rapid technological development, and continuous changes in customer demands (Islam et al. [2006]). To cope with the current business trends, the demands on modern manufacturing systems have required increased flexibility, higher quality standards, and higher innovative capacities (Monica and John [1999]). 'These demands emphasize the need for high levels of overall system reliability that include the reliability of all human elements, machines, equipment, material handling systems and other value added processes and management functions throughout the manufacturing system' (Islam et al. [2006]). Whatever the resources they possess, the manufacturing organizations encounter undesirable events and unwanted setbacks such as machine breakdowns, material shortages, accidents, and absenteeism that make the system unreliable and inconsistent (Monica and John [1999]; Islam [2008]; Islam et al. [2008]; Mitala and Pennathurb [2004]; Monostori et al. [1998]; Toulouse [2002]). In fact, undesirable events and unwanted setbacks (internal and external) in day-to-day operations are common in small and medium-sized manufacturing enterprises (SMEs; Islam [2008]). The authors of this paper chose the word 'disturbance' to represent any of these undesirable events and setbacks. They define the disturbance as 'an undesirable or unplanned event that causes the deviation of system performance in such a way that it incurs a loss,' and the definition is published by the authors elsewhere (Islam et al. [2006]; Islam [2008]). This research adopts the definition of disturbance. As a disturbance creates undesirable consequences that are obviously detrimental to a business performance, we finally refer to a disturbance as a 'risk determinant' on the basis of its significant presence in the system and its consequential negative impact on business and operational performance. Disturbances are linked to undesirable consequences which may originate from different circumstances (Monostori et al.[1998]). 'Whatever the sources of disturbances, the consequences resulting from them could be; difficulties to continue work, decreased productivity, reduced production rate, increased defective products, unplanned rework, delayed delivery to market, unexpected downtime, human loss, etc.' (Islam et al. [2006]; Islam [2008]). In practice, there is a financial loss due to any consequential effects of disturbances. The combined effect of different disturbances could effectively cripple an SME's business performance which may ultimately put it at risk of complete failure (Islam et al.[2006]). The risks can, in general, be categorized into three groups: operational, occupational, or economic. The first category of risks involves the loss of production and the loss of production capability that includes productivity losses, quality-related losses, interrelated activity losses, and asset losses. The second category comprises the risks associated with employees' health, safety, and well-being, while the third category encompasses business risks associated with the financial penalties resulting from either of the first two categories as well as compensation claims and damage to reputation. While dealing with risks, the term 'hazard' automatically comes into the scenario; thus, the definition of a hazard can play an important role when dealing with risks in the industrial context. A hazard is a condition that can cause harm, injury, death, damage, or loss of equipment or personnel (Bahr [1997]) and can exist without anything actually failing within the enterprise. There are four types of hazards, namely *catastrophic* (death or serious personnel injury or loss of a complete system), *critical* (severe injury or loss of valuable equipment), *minor* (minor injury or minor system damage), and *negligible* (no resulting significant injury or system damage). While examining the definitions of a hazard, it can be noticed that a hazard ultimately represents a situation or condition that has the potential to harm people, property, or the environment. However, a question now presents itself, that if there is no chance to harm any of these three elements (people, property, environment), can we classify the situation as a hazard? For an example, the absence of a key machine operator may have no impact on any of these three elements, but it has the potential to develop financial risk to the organization in terms of loss of production; however, the impact might be severe for a small business if the absence is prolonged. There might be some debate as to whether absenteeism should be included in the hazard category or not, but most people would agree to recognize it as a potential operational disturbance which could have serious consequences for an SME. Operational disturbances can be seen from different perspectives and can also be described with various words such as disruptions, failures, errors, defects, losses, and waste (Islam [2008]). However, all potential disturbances and their consequential losses should be considered in the risk management of SMEs because they can be both time-consuming and costly. We believe that this type of disturbance should be studied under the umbrella of risk management. Consequently, while studying risk management in SMEs, we prefer to use the term 'disturbance' instead of hazard. According to our definition, therefore, a disturbance represents all types of hazards as well any other unwanted setback that can produce uncertainty or a loss for an organization. The focus of our research was to identify typical risk determinants of SMEs that need to be considered in developing an integrated risk management approach which should include strategic, operational, occupational, financial, and technology-oriented risks. The research is, therefore, built in a specific research question - what are the typical risk determinants of manufacturing SMEs?

Based on the findings related to the question, we have identified a set of key internal and external operational disturbances, which are eventually highlighted as 'risk determinants' based on their occurrence and consequential effects on the business performance of SMEs. This paper presents the identified risk determinants and describes a methodology to identify them.

SMEs are viewed as a source of flexibility and innovation, and they make significant contributions to the economies of many countries, both in terms of the number of SMEs and the proportion of the labor force employed by them (Hoffman et al. [1998]; Ministry of Economic Development [2004]). However, SMEs are perceived as high-risk ventures, and the entry and exit rates support this perception (Zacharakis et al. [1999]). Previous research has indicated that there is little difference between small business failure rates in developed and developing economies, and it is estimated that 50% of all start-ups fail in their first year, while 75% to 80% fail within the first 3 to 5 years in the USA (Anderson and Dunkelberg [1990]). It has also been shown that up to 50% of the small businesses started in South Africa eventually failed (Watson and Vuuren [2002]). In New Zealand, 40% to 50% of small businesses failed within the first 10 years, and a negative correlation was found between a firm's total full-time employment and its failure rate (Ministry of Economic Development[2004]). Business failure is often caused by a lack of knowledge, misplaced overconfidence, lack of financial performance strategies, or a lack of internal management planning (Gibson and Cassar[2005]; Hartcher et al. [2003]). In spite of high failure rates, however, small businesses continue to be an essential component of the economy of many countries as they account for a significant percentage of all entities and collectively employ large numbers of the workforce. Generally, SMEs depend on financial factors such as profit or sales when considering business risks (Waring and Glendon [1998]). However, monetary factors alone may ignore many issues affecting the long-term reputation of the SME and its staff. A recent research study has suggested that risk management is less well developed within SMEs where the strong enterprise culture sometimes mitigates against managing risks in a professional structured way (Virdi [2005]). According to the study, the SMEs are reluctant to adopt a formal risk management strategy despite having the evidence that businesses that adopt risk management strategies are more likely to survive and grow. Zacharakis et al. ([1999]) identify some reasons for failures of small businesses that include both internal and external causes. The internal causes of failure include poor management, lack of risk management planning, and failure to adopt a risk limit threshold. The external causes included government policies, the vulnerability resulting from small size, competition from larger businesses, civil strife, natural disasters, and general economic downturns. It was also found that 'overconfidence' could often drive small business operators to devalue the importance of fundamental risk assessment that ultimately caused their failure. Although there are some other causes for failure that are highlighted in this section, our research is not intended to investigate the reasons behind the absolute failures of SMEs. Rather, it deals with identifying the potential risks existing when operating SMEs within their current infrastructures so that they can avoid potential failures by implementing a strategic risk management approach. Because manufacturing involves a complicated mix of people, systems, processes, and equipment, an effective research strategy needs to be multidisciplinary in its approach to establishing a risk management framework (Islam [2008]). Because of some infrastructural, technological, financial, and human resource-related limitations, SMEs may keep themselves away from adopting a positive approach towards strategic risk management (Islam et al. [2006]; Islam [2008]; Hartcher et al. [2003]; Martie-Louise [2006]). Islam et al. ([2006]) state:

It is noteworthy to mention that major accidents and emergencies rarely occur in SMEs although small losses, near misses, unsafe acts and unsafe conditions are common occurrences. But, problems, failures and mistakes as well as incorrect or ineffective actions, are very likely occurrences in the daily business of SMEs and for this reason, in practice, minor incidents and near misses are worth analyzing since in slightly different circumstances the consequences could have been quite serious. By monitoring even small problems and analyzing their underlying causes, it might be possible to discover causes for more serious problems and the existence of hazards. Therefore, no disturbance should be overlooked or should be allowed to happen again.

In the authors' knowledge, research works done on risk management have generally focused on particular industries such as nuclear, aviation, space exploration, chemical processing, and other areas where the consequence of a system breakdown is considered severe or catastrophic for human beings or the environment, and/or where the potential financial loss is significant (Islam et al.[2006]; Andrews and Moss [2002]; Khan and Abbasi [1998]; Milan [2000]; Seastroma et al. [2004]; Strupczewski [2003]). In addition, research works on risk management in other areas, including financial sectors, medical science, transportation, and construction engineering, have also significantly expanded with time (Islam et al. [2006]). In contrast to this, lower priority has been noticed in the literature concerning risk management in the SME sector. Most of the studies relevant to risk management in this sector indeed concentrate solely on the risks associated with safety and occupational health (Islam et al. [2006]; Islam [2008]). Protective practices such as occupational safety and health and other safety-related programs should, if properly implemented and practiced, ensure better health and working environments inside organizations. They do not, however, ensure the smooth running of the organization or minimize its risks operationally, technically, and/or financially.

Hazard identification within a system is the starting point of any risk identification or assessment process that emphasizes the critical components or factors that produce or could produce failure or harmful consequences for humans, assets, or the environment (Islam et al. [2006]; Islam [2008]). In this context, different techniques such as Hazard and Operability Analysis studies, Failure Mode and Effect Analysis, Failure Mode and Effect Critical Analysis, Hazard Analysis with Critical Control Points, Fault Tree Analysis, Event Tree Analysis, 'What if' analysis, and Checklists are widely used in practice (Islam et al. [2006]; Khan and Abbasi [1998]; Mushtaq and Chung [2000]; Pearson and Dutson[1995]; Tixier et al. [2002]). All these techniques focus on the main hazard sources systematically, but none of them can produce a thorough list of important system failures, causes, consequences, and controls and can lend themselves to rigorous risk acceptability analysis (Islam et al. [2006]). Furthermore, none of the techniques are necessarily effective in identifying and prioritizing the risks associated with multifaceted criteria. None of the abovementioned methods alone can readily be applicable for dealing with risks associated with operational disturbances, because of their complex nature. 'For example, a disturbance such as 'tool shortage' could be rooted in; erroneous planning of stock, misuse by the operator, unexpected breakage, or incorrect selection of tool for the particular task. Thus, the origin of the disturbance could either be strategic, operational or technical. This means that a detailed analysis of a particular disturbance is required to establish a suitable risk handling procedure' (Islam et al. [2006]). In this connection, we have developed a strategic risk management model for SMEs and have published the model elsewhere (Islam et al. [2006]; Islam et al. [2008]). However, we conducted further study on the identification of specific risk determinants of SMEs and have discussed the identified determinates in this paper.

CASE DESCRIPTION

Research Methodology

We choose an empirical investigation as it puts special emphasis on the affiliated research leading to the development of a strategic risk management framework in terms of operational and organizational aspects (Islam [2008]; Glaser and Strauss [1980]; Luis et al. [1999]; Mills et al. [1995]; Pettigrew et al. [1989]). The empirical investigation was carried out by applying a multi-method approach (combination of case study and survey methods), called triangulation, which provided a relatively potent means of assessing the degree of convergence, as well as identifying divergences, between the results obtained (Islam [2008]; Brewer and Hunter [2006]; Jick [1979]). In the triangulation method, the survey results improved the authors' understanding of the particular phenomenon (relationship between potential disturbances and their associated risks in this case). On the other hand, the case studies added to a more holistic and richer contextual understanding of the survey results. Thus, the multi-method approach is believed to be enhancing the credibility of the research results while reducing the risk of observations reflecting some unique artifact (Brewer and Hunter [2006]; Denzin [1989]).

Data Collection Methods and Sample

For the empirical investigation, standard questionnaires were developed and verified by a panel of academic experts and subsequently by an industry focus group in a pilot study. The questionnaires were designed to explore the risk determinants (potential disturbances) and risk indicators (detrimental parameters to business performance) relating to existing practices in the studied organizations. The focal points of the questionnaire were (1) production-related activities associated with risks; (2) quality, reliability, and health- and safety-related issues of both assets and personnel; (3) major activities in the supply chain networks; and finally, (4) strategic issues relating to the current practices in risk management.

There were two phases in the data collection process. In the first phase, questionnaires were sent to 55 manufacturing SMEs (to 165 individual management personnel, to three tiers of management of each organization), and in the second, to 157 SMEs (to 417 management personnel). The respondents were given 1 month to return the completed questionnaires while an additional 3 weeks were allocated for telephoning and personal interviewing to acquire missing data in incomplete questionnaires. Out of 212 SMEs, 11 SMEs declined to participate in the questionnaire survey due to their organizational restructuring, busy scheduling of the management, absorption in other business sectors, or some other undisclosed reasons, though they mentioned their keen interest (in the response letters) to the research subject. Four sets of questionnaires were sent back to the researchers not finding the addressee. Five participating organizations provided partially completed questionnaires, which have been excluded in the analysis. Altogether, 96 usable responses from management personnel (top, middle, and front-line management), from 32 responding SMEs, were returned and have been analyzed, and presented in this paper. It is noted that the organization which returned three sets of completed questionnaire is only considered as responding SME. In this connection, the useful response rate of 18.27% from companies was considered satisfactory and representative of SMEs in New Zealand. The overall response rate of 23.08% from the selected SMEs indicates the substantial importance of the research topic, while past experience suggests that mail survey response rates are often low and appear to be declining among small business populations (Dennis [2003]). However, before making any conclusive remarks on the survey findings, further verification was carried out by subsequent in-depth case studies involving five SMEs from among the participants in the mail survey. We choose the follow-up case study approach as '… an empirical inquiry that investigates a contemporary phenomenon within some real-life context and a methodology involving multiple sources of data which provides the fullest understanding of the phenomenon and improves the validity of research implications through triangulation' (Scudder and Hill [1998]; Yin [1994]).

Validation of Questionnaire

The mailed survey was carried out by the developed questionnaires. One questionnaire was designed for top management (senior executives), and the other was designed for middle and front-line management of each organization. The purpose of two separate questionnaires was to collect disturbance information from different areas of concerns of each management level. In total, 26 questions were formulated for the questionnaire of the top management and 34 questions were in the questionnaire for middle management. However, in the context of this paper, the questions that were directly related to the disturbances are presented in 1 for the clarity of the investigation. Most of these questions were of the 'multiple choice' kind. The answers of the questions comprised four-point rating scales for response. The four-point rating scale was chosen to prevent the occurrence of central tendency error.

A typical example of the questions related to an internal disturbance is, Over the last 12 months, how often did you notice 'absenteeism' in your organization? (4 = often, 3 = sometimes, 2 = rarely, and 1 = never). A typical example of the questions related to an external disturbance is, To what extent does 'skilled labor shortage' impede your business performance (profit/growth)? (4 = to great extent, 3 = to some extent, 2 = a little amount, and 1 = not at all). A typical example of the questions related to a risk indicator is, Over the last 12 months, how often did you notice 'lower than expected productivity'? (4 = often, 3 = sometimes, 2 = rarely, and 1 = never).

The questionnaires were designed in such a way that they were easy to understand and answer. They were pretested and carried out in two sequential stages. The first stage consisted of a review by a panel of academic experts and survey specialists who ensured that all necessary questions were included and ambiguous questions eliminated, and the categorization of the questions was set up properly to ensure that subsequent data analysis would provide the desired information. The second stage was a pilot study with ten participating SMEs. The responses from the pilot study allowed the authors to verify whether respondents were biased towards certain categories of questions or leaving questions unanswered. The study found that all respondents answered all questions and the responses on the ordinal scales were reasonably dispersed. Finally, the measuring scales were tested to verify the reliability of instrument with the help of Cronbach's alpha (α) (Hinton [2004]; Black[1999]). The values of α were 0.701 and 0.716, and 0.721 for the questions of internal and external disturbances, and risk indicators (consequential effect), respectively, that ensured the reliability and internal consistency of the measuring scales.

Characteristics of Studied SMEs

The significance of the SME sector in New Zealand has been increasing, with further opportunities presented by globalization and technological development (Ministry of Economic Development[2004]). New Zealand is a small nation state of 4.3 million people, ethnically diverse, with a strong culture of self-help and independence underpinning business development (Ministry of Economic Development [2004]). New Zealand's size means that by international standards, its small businesses are very small but are the dominant sector in terms of employment, organizational structure, and social and economic cohesion. A recent report on SMEs states that in the context of policy consideration, the characteristics of small-sized businesses should typically include personal ownership and management, few specialist managerial staff, and not being part of a larger business enterprise (Ministry of Economic Development [2003]). SMEs in New Zealand typically exhibit these characteristics, and it is in this context that our research has been designed to deal with companies with employment in the range of 10 to 100 employees (Islam [2008]).

The list of SMEs selected for the mail survey and case studies was compiled from a variety of business databases; these were randomly chosen to represent a range of manufacturing groups. These groups covered the four sectors of (1) metal-based product and equipment manufacturers, (2) wood and wood-based product manufacturers, (3) paper- and plastic-based product manufacturers, and (4) textile and garment manufacturers. These groups were selected because of their economic importance to New Zealand. The characteristics of the participating SMEs in the mail survey are presented in Table 1.

Table 1: Characteristics of the selected SMEs

Classification	Criteria	Number of organizations	Percentage of organizations (%)
Firm size	Small size (10 to 25 employees)	12	37.50
	Medium size (26 to 100 employees)	20	62.50
Annual turnover (New Zealand $)	Less than 5 million	4	12.50
	Between 5 and 25 million	20	62.50
	Between 25 and 50 million	6	18.75
	Over 50 million	2	6.25
Business category	Metal-based product and machinery manufacturing	16	50.00
	Textile and garment manufacturing	6	18.75
	Wood-based product and furniture manufacturing	6	18.75
	Plastic- and paper-based product manufacturing	4	12.50
Plant set-up	Single site	20	62.50
	Multi-domestic sites	9	28.13
	Multinational sites	3	9.38
Employment contracts	Nil	11	34.38
	Less than 5% of total employees	17	53.13
	Between 5% and 10% of total employees	2	6.25
	More than 10% of total employees	2	6.25
Total number of selected organizations		*32*	*100.00*

Islam and Tedford

Islam and Tedford Journal of Industrial Engineering International
2012 8:12, doi: 10.1186/2251-712X-8-12

Key Findings and Analysis

The key findings are categorized and presented in the following sections:

Risk Indicators

Two principal measures of corporate performance are profit rate and growth rate (Freel [2000]; Geroski and Machin [1992]; Wynarczyk and Thwaites [1996]). Needless to say, there are a number of ways to measure growth rate and profitability which are substantially linked to several variables of operational activities. Several studies have overwhelmingly indicated that effective employee management, along with other strategic measures, can lead to a competitive advantage in the form of a motivated workforce, improved operational and business performance, reduced employee turnover, and improved productivity, which in turn improve the net profit of a firm (Batt [2002]; Macduffie [1995]; Virdi [2005]). Moreover, growth of a business would appear to play an important role in its sustainability in a dynamic business (Barbara et al. [2000]). We could, therefore, interpret that dissatisfaction with net profit and in business growth (assuming that the business plan is realistic), as well as significant employee turnover rates, could be the results of inappropriate or inadequate strategic allocation and utilization of resources and that these should be treated as primary indicators of potential problems for an organization. Our research approach, however, was not to verify the measures of these categories. Rather, it tried to identify whether there is any correlation between business growth rate and net profit, and the potential disturbances. The research finds that approximately 32% of the SMEs are dissatisfied with their existing 'net profit' (of which 10% are very dissatisfied) and about 40% are dissatisfied with 'business growth' (of which 10% are very dissatisfied). On the other hand, 9% of the organizations are very satisfied with both net profit and business growth. The study also finds that 30% of SMEs consider the existing 'employee turnover rate' as a substantial impediment to effective business operation, while 43% indicate the impediment from this factor to be small, and 26% indicate it to be negligible. These are apparently linked to operational risks of direct or indirect losses due to failures in systems, processes, and people or from external factors. Thus, dissatisfaction level with net profit and in business growth and employee turnover rate is considered as 'risk indicators' for our research. In addition to these three, 11 risk indicators which are linked to operational, occupational, and economic losses are identified from the study.Figure 1 shows the relative position of these risk indicators in terms of their emergence in the systems of the studied SMEs.

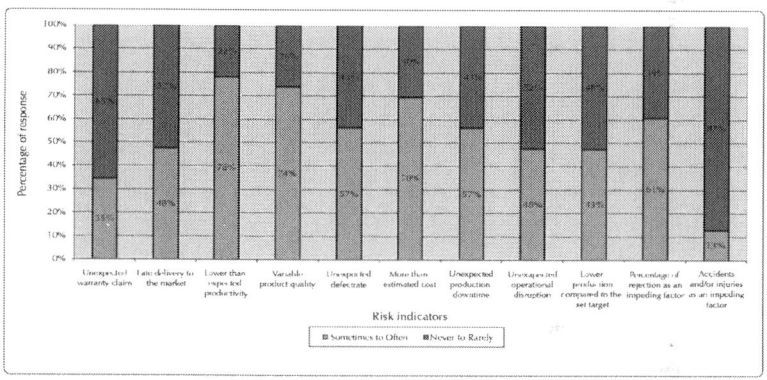

Figure 1: Status of the potential risk indicators in terms of their presence in the studied SMEs.

Operational Disturbances

The risk indicators have potential linkages with day-to-day operational disturbances, which degrade business performance and the business environment. In consequence, the disturbances ultimately play a vital role in putting an organization at risk in terms of production, safety, and financial, resulting from both internal and external customer dissatisfaction (Islam [2008]). These can lead to a loss of market share and eventually put the *organization* out of business, if they are not carefully treated. For this, a thorough investigation was conducted to identify key operational disturbances (in essence, driving risk factors) and their linkage to some risk indicators discussed in the previous section. We have identified a number of notable internal and external operational disturbances, which are summarized in Figures 2 and 3.

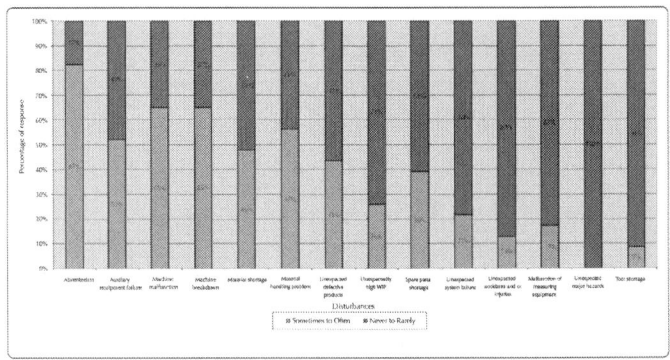

Figure 2: Status of the potential internal disturbances in the studied SMEs.

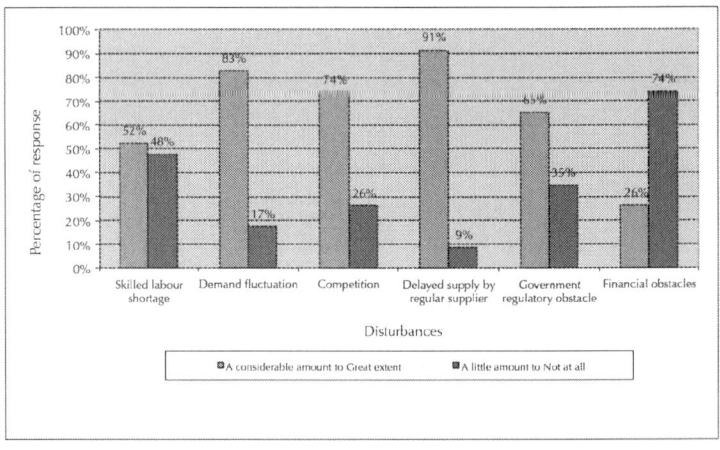

Figure 3: Status of the potential external disturbances in the studied SMEs.

Among the internal disturbances, *absenteeism*, *machine malfunction*, *machine breakdown*, and *material handling disruption* were found to be the most significant disturbances, and *unexpected major hazards, unexpected accidents/injuries,* and *tool shortage* were found to be the least significant ones, while the other disturbances were found to fall between these extremes. Among the external disturbances, *competition, delayed supply by the regular supplier,* and *skilled labor shortage* were found to be the most significant ones, while *financial obstacle* was found to be the least significant in terms of their influence on the operational system. However, despite the minimal influence of some disturbances, they were still considered for further analysis to find out their consequential effects.

Risk Determinants

All disturbances presented in Figures 2 and 3 were considered for further analysis to determine whether they should be treated as risk determinants. The analysis included some statistical methods of parametric and non-parametric testing such as *t test*the *Friedman test,* and the *Spearman correlation coefficient tests* (Hinton [2004]) at two significant levels: $= 0.01$ (99% confidence level) and $= 0.05$ (95% confidence level). The results of the *t* test are presented in Table 2.

Table 2: Results of t tests with internal disturbances

	d2	d3	d4	d5	d6	d7	d8	d9	d10	d11	d12	d13	d14
Absenteeism (d1)	3.213**	3.026**	2.621**	3.346**	5.724**	2.421*	3.471**	4.592**	7.854**	5.738**	6.098**	11.754**	10.817**
Auxiliary equipment failure (d2)		-0.264	0.190	0.711	0.846	-0.549	1.283	1.899*	2.998**	3.696**	3.581**	6.789**	6.260**
Machine malfunctions (d3)			0.374	1.121	1.371	-0.264	1.513	2.197**	3.899**	4.243**	6.244**	7.414**	7.468**
Machine breakdown (d4)				0.514	0.753	-0.537	0.983	1.274	2.913**	2.731**	3.366**	6.249**	4.136**
Material shortage (d5)					0.309	-1.371	0.989	1.429	2.584**	3.931**	3.845**	6.969**	5.359**
Material handling problem (d6)						-1.429	0.789	1.045	2.989**	2.532**	3.581**	5.778**	6.054**
Unexpected defective product (d7)							3.308**	3.283**	3.329**	5.381**	4.333*	6.875**	8.125**
Unexpected work-in-progress (d8)								0.000	0.560	1.077	1.663	2.278**	3.696**

Spare parts shortage (d9)	0.783	1.899*	2.062*	2.954**	6.278**						
Unexpected system failure (d10)		0.437	1.435	6.696**	3.280**						
Unexpected accidents/ injuries (d11)			1.208	2.249*	3.638**						
Malfunctions of measuring equipment (d12)				0.711	1.986*						
Unexpected major hazard (d13)					1.295						
Tool shortage (d14)											

Numbers are the values of t's; *p < 0.05; **p < 0.01.

Islam and Tedford

Islam and Tedford Journal of Industrial Engineering International 2012 8:12, doi: 10.1186/2251-712X-8-12

On the basis of their comparative occurrence in practice, the disturbances are assigned with relative scores. The disturbance which occurs most frequently is assigned with the highest score, while the disturbance which occurs least frequently is assigned with the lowest score. Thus, among the internal disturbances, 'absenteeism' scores the highest number of points, and 'tool shortage' and 'unexpected major hazard' jointly score the lowest.

The relative positions of the internal disturbances, based on their scores, are shown in the second column of Table 3. The final test results (based on Spearman's correlation coefficient, r_s) confirm the positive correlation between internal disturbances and risk indicators; the results are presented in Table 4. Based on the positive correlation of disturbances with a number of risk indicators, scoring is performed. The highest scorer is correlated with a maximum number of risk indicators, while the lowest one is correlated with a minimum number of risk indicators. Thus, all disturbances are assigned with scores and are presented in the third column of Table 3. Finally, on the basis of the product of two scores (one for appearance or occurrence and the other for correlation), final ranking is performed for the risk determinants. The determinant which scores the maximum value is assigned with the highest rank (1), and the determinant which scores the minimum value is assigned with the lowest rank (14). Accordingly, the relative ranking for all risk determinants is established and is shown in the fifth column of Table 3. According to the final ranking, 'absenteeism' becomes the most important (number 1) risk determinant among the internal disturbances, while 'malfunctions of measuring equipment' becomes the least important one. Similar tests were conducted and relative measures were performed on the external disturbances, the results of which are summarized in the second and third columns of Table 5.

Table 3: Potential risk determinants (internal)

Risk determinants	Scores of the disturbances based on the distribution of the frequency of occurrence (F)	Scores of the disturbances based on their positive correlation with the risk indicators (C)	Total score	Final ranks of the risk determinants
	(14 = highest score, 1 = lowest score)	(14 = highest score, 1 = lowest score)	(F × C)	(1 = most important, 14 = least important)
Absenteeism	14	14	196	1
Unexpected defective product	13	9	117	2
Machine malfunctions	12	5	60	5
Auxiliary equipment failure	11	4	44	7.5
Material shortage	9	12	108	3
Material handling problem	9	11	99	4
Machine breakdown	9	6.5	58.5	6
Spare parts shortage	7	3	21	10
Unexpected work-in-progress	5.5	2	11	12
Unexpected system failure	5.5	8	44	7.5
Unexpected accidents or injuries	4	10	40	9
Malfunctions of measuring equipment	3	1	3	14
Unexpected major hazard	1.5	6.5	9.75	13
Tool shortage	1.5	13	19.5	11

Table 4: Correlation between internal disturbances and risk indicators

	D1	D2	D3	D4	D5	D6	D7	D8	D9	D10	D11	D12	D13	D14
Absenteeism	0.67**	IPC	0.39*	IPC	0.77**	IPC	0.30*	0.33*	0.40*	0.32*	IPC	0.48**	0.64**	0.60**
Auxiliary equipment failure	0.49*	IPC	IPC	0.47*	0.45**	0.50**	IPC	NC	0.31*	NC	NC	IPC	0.36*	IPC
Machine malfunctions	0.44*	IPC	IPC	0.33*	IPC	0.43*	NC	0.30*	IPC	NC	0.40*	IPC	0.41*	0.39*
Machine breakdown	0.42*	IPC	0.44*	IPC	0.40*	IPC	0.30*	IPC	0.55**	0.35*	0.51**	NC	IPC	IPC
Material shortage	0.54**	0.62**	0.33*	0.40*	0.38*	NC	0.53**	0.36*	0.51**	0.35*	IPC	IPC	NC	NC
Material handling problem	0.49**	0.53**	0.39*	0.44*	0.61**	IPC	IPC	0.33*	IPC	0.49**	0.43*	IPC	IPC	IPC
Unexpected defective product	0.36*	IPC	IPC	IPC	0.52**	NC	0.47**	NC	NC	0.56**	NC	0.33**	0.35*	0.45**
Unexpected work-in-progress	IPC	NC	IPC	NC	IPC	IPC	NC	0.41*	0.33*	IPC	NC	0.45**	IPC	NC
Spare parts shortage	0.35*	NC	0.38*	IPC	0.35*	IPC	IPC	NC	0.51**	NC	NC	0.34*	0.38*	IPC
Unexpected system failure	0.39*	IPC	0.43*	IPC	0.53**	IPC	0.36*	NC	NC	0.57**	0.40*	NC	IPC	0.45**

	D1	D2	D3	D4	D5	D6	D7	D8	D9	D10	D11	D12	D13	D14
Unexpected accidents/injuries	0.04*	IPC	0.34*	0.37*	IPC	0.38*	IPC	0.43*	1.00**	NC	0.41*	0.44*	NC	NC
Malfunctions of measuring equipment	IPC	IPC	NC	NC	NC	NC	NC	0.50**	IPC	NC	0.52**	IPC	IPC	NC
Unexpected major hazard	0.39*	IPC	0.43*	IPC	0.49**	0.32*	0.32*	NC	0.31*	0.45**	0.41*	NC	IPC	NC
Tool shortage	0.48*	IPC	0.34*	0.37*	0.43*	0.50**	IPC	0.33*	0.37*	NC	IPC	0.49**	0.73**	0.45**

Numbers in the boxes are the values of r_s; IPC, insignificant positive correlation; NC, no correlation; D1, lower than expected productivity; D2, variable product quality; D3, unexpected defect rate; D4, more than estimated cost; D5, unexpected production downtime; D6, unexpected operational disruption; D7, lower production compared to the set target; D8, percentage of rejection at various levels; D9, accidents and/or injuries; D10, late delivery to the market; D11, unexpected warranty claim; D12, employee turnover rate; D13, dissatisfaction level with net profit; D14, dissatisfaction level with business growth; $*p < 0.05$; $**p < 0.01$.

Islam and Tedford

Islam and Tedford Journal of Industrial Engineering International 2012 8:12, doi: 10.1186/2251-712X-8-12

Table 5: Potential risk determinants (external)

Risk determinants	Scores of the disturbances based on the distributions of the level of impediments on business (i)	Scores of the disturbances based on their positive correlation with the risk indicators (c)	Total score	Final ranking of the risk determinants
	(6 = highest score, 1 = lowest score)	(6 = highest score, 1 = lowest score)	(i × c)	(1 = most important, 6 = least important)
Delayed supply by the suppliers	6	2.5	15	2
Demand fluctuation	5	2.5	12.5	4
Competition	4	4.5	18	1
Skilled labor shortage	3	4.5	13.5	3
Government regulations	2	1	2	6
Financial obstacles	1	6	6	5

Islam and Tedford

Islam and Tedford Journal of Industrial Engineering International 2012 8:12, doi: 10.1186/2251-712X-8-12

Discussion and Evaluation

The findings from the mail survey have been presented in the previous section. Most of the findings have strongly been supported by the findings from case studies. Both investigations confirm that there are some typical internal and external operational disturbances, which expose SMEs to operational risks. Comparative findings from the two investigations are depicted in Figures 4 and 5. The comparison for disturbances is made on an extended scale of 1 to 10 in terms of their frequency of occurrence (for internal disturbances) and of their detrimental effects on operational performance (for external disturbances). Figure 4 shows that both investigations identify 'absenteeism' as the most frequently occurring internal disturbance and 'tool shortage' as the least frequently occurring in the SMEs studied, while the others fall between these two extremes. Figure 5 shows that 'delayed supply by regular suppliers' (very closely followed by 'demand fluctuation' and 'competition') is the most detrimental external disturbance, and 'financial obstacles' is the least detrimental to the SMEs. Both investigations further confirmed a set of risk indicators, which can be used as the consequential effects resulting from the disturbances (Figure 6). These risk indicators are linked to operational, occupational, and economic losses. The findings of both investigations again converge on the same conclusions, in terms of the overall ranking of the disturbances, even though there are slight, statistically insignificant variations in some cases.

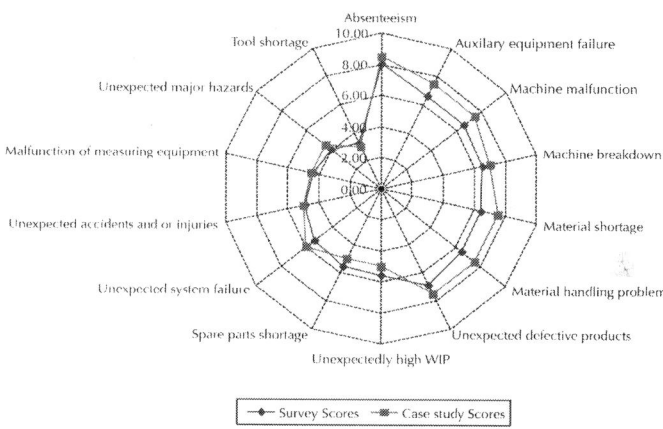

Figure 4: Comparative images between the survey and case study results on internal disturbances.

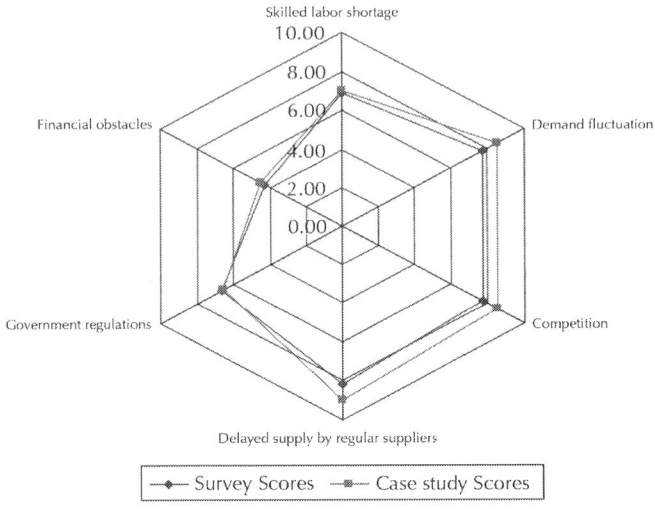

Figure 5: Comparative images between the survey and case study results on external disturbances.

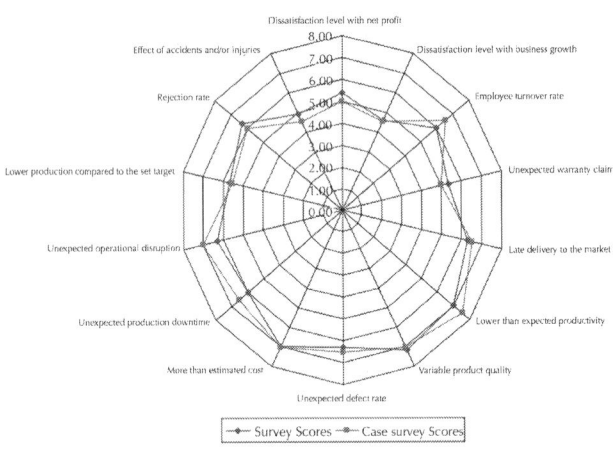

Figure 6: Comparative images between the survey and case study results on risk indicators.

The research study reveals that SMEs have, in general, inadequate measures and planned strategies in place to deal with such risk determinants. Thus, the identified set of internal and external risk determinants found from this study will play a vital role in ensuring that SMEs realize the strengths and weaknesses in their ability to cope with the identified internal factors, as well as the threats and opportunities arising from the identified external factors, while assisting them in formulating and implementing strategic measures to deal with the resulting operational risks. It is obvious that some disturbances are more detrimental than others. Moreover, the nature of the disturbance is found to be dynamic and idiosyncratic in nature. The dynamic behavior of a disturbance in different organizational settings and in different time frame leads us to a common understanding that the appearance of a particular disturbance varies from organization to organization and time to time. Moreover, the same disturbance produces different consequential effects to different organizations based on its time of occurrence and the duration of its existence in the system. An organization, therefore, needs to identify the characteristics of the various disturbances and their consequential effects over time, to develop a proactive strategy for managing operational risks.

CONCLUSIONS

An organization is basically a giant network of interconnected nodes. Changes in one part of an organization can affect other parts of the organization with surprising and often negative consequences. The minimization of delays in the system generally becomes an important issue in lean manufacturing. In this context, the optimization of response time to changes in the external environment becomes vital. At the same time, smooth and consistent operational performance in the internal environment is necessary to continue the business in this dynamic business world. Internal and external disturbances to its day-to-day operation put an SME at risk in terms of production, safety, and the business itself. The risks associated with disturbances can be detected by analyzing the negative or detrimental consequential effects, which are identified as risk indicators in the research. We have identified some typical internal and external operational disturbances that need to be considered as risk determinants for SMEs. It is found that some disturbances are positively correlated with a greater number of risk indicators and some with a lesser number of indicators. It is also found that every disturbance is significantly correlated with at least one of the risk indicators. This means that in terms of operational risks, an SME needs to consider all the identified disturbances (risk determinants) in its strategic decisions for managing operational risks successfully.

We find that the majority of the studied SMEs do not have systematic risk management strategies in place. It is discovered that the majority of SMEs used standard hazard identification forms, which comply with the requirements of the Health and Safety in Employment Act in New Zealand (Avery[1993]). The current practices in SMEs regarding risk identification relies, almost exclusively, on the documented records of industrial injuries which these forms produce. Near misses are not generally recorded even though this is a requirement of the legislation. Moreover, the identification of root causes of the risk determinants and their related origins is not practiced in the studied SMEs, and in SMEs where it is practiced to some extent, the flow of information tends to miss many of the relevant personnel. In addition, the disturbance handling systems in these organizations, in terms of data collection, information processing, information sharing, and decision making, are found to be relatively weak and very informal. With regard to the identification of external disturbances, most SMEs do not have assessment criteria in place to measure the consequences, nor have enough information available to help them determine their root causes.

The identified set of internal and external risk determinants should provide a quick reference or benchmark for SMEs. The struggle with the identification of operational risk determinants should be minimized by the identified set of determinants, obtained from a representative sample of SMEs in New Zealand. It is, however, relevant to note that the relative rankings of the identified risk determinants could vary from organization to organization based on their likelihood of occurrence and their impact on business performance. The individual business setting, including current strategic measures, practices, and vulnerability, would play a vital role in developing appropriate strategic plans and actions in each case. While it may be necessary for organizations to add or delete determinants to those identified in this research, depending on their particular situation, they should be able to apply the described methodology to assist them in identifying the risk determinants appropriate to them. In this way, they should be able to identify the extent of the risks associated with the determinants by incorporating the metrics of time, money, and asset loss due to these. In conclusion, the research findings presented in this paper will, hopefully, add to the body of knowledge on good practices in risk management resulting from operational disturbances which can affect SMEs and that may also be useful to both management professionals and researchers in the field of risk management.

AUTHORS' CONTRIBUTIONS

Dr. MAI designed the research, collected and analyzed the data, and drafted the manuscript. Dr. DT substantially contributed to the conception and design phase, modification of the questionnaires and analysis, and editing of the manuscript critically for its intellectual content. Both authors read and approved the final manuscript.

REFERENCES

1. Anderson RL, Dunkelberg JS (1990) Entrepreneurship: starting a new business. Harper and Row, New York.

2. Andrews JD, Moss TR (2002) Reliability and risk assessment. Professional Engineering, London.

3. Avery M (1993) Health & safety laws at work: key issues. Teemay Consultants, New Zealand.

4. Bahr NJ (1997) System safety engineering and risk assessment: a practical approach. Taylor & Francis, Washington, DC.

5. Barbara JO, Sandy HS, Allan LR (2000) Performance, firm size, and management problem solving. Journal of Small Business Management 38(4):42-58

6. Batt R (2002) Managing customer services: human resource practices, quit rates, and sales growth. Academy of Management Journal 45:587-597

7. Black TR (1999) Doing quantitative research in the social sciences: an integrated approach to research design, measurement and statistics. Sage, Thousand Oaks.

8. Bowman C, Ambrosini V (1997) Using single respondents in strategy research. British Journal of Management 8:119-131

9. Brewer J, Hunter A (2006) Foundation of multi-method research: synthesizing styles. Sage, Thousand Oaks.

10. Charmaz K (2006) Constructing grounded theory - a practical guide through qualitative analysis. Sage, Thousand Oaks.

11. Dennis WJ (2003) Raising response rates in mail surveys of small business owners results of an experiment. Journal of Small Business Management 41(3):287-295

12. Denzin NK (1989) The research act: a theoretical introduction to sociological method. Prentice Hall, Englewood Cliffs.

13. Freel MS (2000) Do small innovating firms outperform non-innovators? Small Business Economics 14(3):195-210

14. Geroski P, Machin S (1992) Do innovating firms outperform non-innovators? Business Strategy Review Summer 3:79-90

15. Gibson B, Cassar G (2005) Longitudinal analysis of relationships between planning and performance in small firms. Small Business Economics 25(3):207-222

16. Glaser BG, Strauss AL (1980) The discovery of grounded theory - strategies for qualitative research, 11th printing. Aldine, New York.

17. Hartcher J, Allan H, Scott H (2003) Perceptions of risks and risk management in small firms. Small Enterprise Research: The Journal of SEAANZ 11(2):71-92

18. Hinton PR (2004) Statistics explained. Routledge, New York.

19. Hoffman K, Milady P, Bessant J, Perren L (1998) Small firms' R&D, technology and innovation in the UK: a literature review. Technovation 18(1):39-55

20. Islam MA (2008) Risk management in small and medium-sized manufacturing organization in New Zealand. The University of Auckland,

21. Islam MA, Tedford JD, Haemmerle E (2006) Proceedings of the 2006 IEEE International Conference on Management and Innovation and Technology, Singapore. In: Strategic risk management approach for small and medium-sized manufacturing enterprises (SMEs)—a theoretical framework. Singapore, IEEE. pp 694-694

22. Islam MA, Tedford JD, Haemmerle E (2008) Managing operational risks in small- and medium-sized enterprises (SMEs) engaged in manufacturing–an integrated approach. International Journal of Technology, Policy and Management 8(4):420-441

23. Jeynes J (2002) Risk management: 10 principles. Butterworth-Heinemann, Oxford.

24. Jick TD (1979) Mixing qualitative and quantitative methods: triangulation in action. Administrative Science Quarterly 24:602-610

25. Khan FI, Abbasi SA (1998) Techniques and methodologies for risk analysis in chemical process industries. Journal of Loss Prevention in the Process Industries 11:261-277

26. Luis EQ, Felisa MC, Serge W, Christopher OB (1999) A methodology for formulating a business strategy in manufacturing firms. International Journal of Production Economics 60–61:87-94

27. Macduffie JP (1995) Human resource bundles and manufacturing performance: organizational logic and flexible production systems in the world auto industry. Industrial and Labor Relations Review 48:197-221

28. Martie-Louise V (2006) Strategy-making process and firm performance in small firms. Journal of Management and Organization 12:209-222

29. Milan J (2000) An assessment of risk and safety in civil aviation. Journal of Air Transport Management 6:43-50

30. Mills J, Platts K, Gregory M (1995) A framework for the design of manufacturing strategy process: a contingency approach. International Journal of Operations and Production Management 15(4):17-49

31. Ministry of Economic Development (2003) SMEs in New Zealand: structure and dynamics. Ministry of Economic Development, Wellington.

32. Ministry of Economic Development (2004) SMEs in New Zealand: structure and dynamics. Ministry of Economic Development and Statistics New Zealand, Wellington.

33. Mitala A, Pennathurb A (2004) Advanced technologies and humans in manufacturing workplaces: an interdependent relationship. International Journal of Industrial Ergonomic 33:295-313

34. Monica PB, John RW (1999) HEDOMS—human errors and disturbance occurrence in manufacturing systems: toward the development of an analytical framework. Human Factors and Ergonomics in Manufacturing 9(1):87-104

35. Monostori L, Szelke E, Kadar B (1998) Management of changes and disturbances in manufacturing systems. Annual Reviews in Control 22:85-97

36. Mushtaq F, Chung PWH (2000) A systematic Hazop procedure for batch processes, and its application to pipeless plants. Journal of Loss Prevention in the Process Industries 13:41-48

37. Pearson AM, Dutson TR (1995) HACCP in meat, poultry and fish processing. Blackie Academic and Professional, New York.

38. Pettigrew AM, Whipp R, Rosenfeld R (1989) Competitiveness and the management of strategic change process: a research agenda. In: Francis A, Tharakan M (eds) The competitiveness of

European industry: country, policies and company strategies, Routledge, London. p 36

39. Scudder GD, Hill CA (1998) A review and classification of empirical research in operations management. Journal of Operations Management 16:91-101

40. Seastroma JW, Peercy RL, Johnson GW, Sotnikov BJ, Brukhanov N (2004) Risk management in international manned space program operations. Acta Astronautica 54:273-279

41. Strupczewski A (2003) Accident risks in nuclear-power plants. Applied Energy 75:79-86

42. Tixier J, Dusserre G, Salvi O, Gaston D (2002) Review of 62 risk analysis methodologies of industrial plants. Journal of Loss Prevention in the Process Industries 15:291-303

43. Toulouse G (2002) Accident risks in disturbance recovery in an automated batch-production system. Human Factors and Ergonomics in Manufacturing 12(4):383-406

44. Virdi AA (2005) Risk management among SMEs–executive report of discovery research. The Consultation and Research Centre of the Institute of Chartered Accountants in England and Wales, London.

45. Waring A, Glendon AI (1998) Managing risk: critical issues for survival and success into the 21st century. International Thomson Business, London.

46. Watson ML, Vuuren JJ (2002) Entrepreneurship training for emerging SMEs in South Africa. Journal of Small Business Management 40(2):154-161

47. Wynarczyk P, Thwaites A (1996) The financial performance of innovative small firms in the UK. In: Oakey R (ed) New technology based firms in the 1990s, 1st edn. Paul Chapman, London.

48. Yin RK (1994) Case study research. Sage, London.

49. Zacharakis AL, Meyer GD, DeCastro J (1999) Differing perceptions of new venture failure: a matched exploratory study of venture capitalists and entrepreneurs. Journal of Small Business Management 37(3):1-14

Citations

CHAPTER 1

Julia Di Domenico, Carlos André Vaz Jr., Maurício Bezerra de Souza Jr., Quantitative risk assessment integrated with process simulator for a new technology of methanol production plant using recycled CO2, Journal of Hazardous Materials, Volume 274, 15 June 2014, Pages 164-172, ISSN 0304-3894, http://dx.doi.org/10.1016/j.jhazmat.2014.02.045.

CHAPTER 2

Chinedu I. Ossai, "Advances in Asset Management Techniques: An Overview of Corrosion Mechanisms and Mitigation Strategies for Oil and Gas Pipelines," ISRN Corrosion, vol. 2012, Article ID 570143, 10 pages, 2012. doi:10.5402/2012/570143.

CHAPTER 3

Rolf K. Eckhoff, "Dust Explosion Prevention and Mitigation, Status and Developments in Basic Knowledge and in Practical Application," International Journal of Chemical Engineering, vol. 2009, Article ID 569825, 12 pages, 2009. doi:10.1155/2009/569825.

CHAPTER 4

Jonatan Gehandler, Road Tunnel Fire Safety and Risk: a Review, doi: 10.1186/s40038-015-0006-6.

CHAPTER 5

Kris Lawry and Dirk John Pons, "Integrative Approach to the Plant Commissioning Process," Journal of Industrial Engineering, vol. 2013, Article ID 572072, 12 pages, 2013. doi:10.1155/2013/572072.

CHAPTER 6

Ana Paula Barquet, Henrique Rozenfeld, and Fernando A Forcellini, An Integrated Approach to Remanufacturing: Model of a Remanufacturing System, doi: 10.1186/2210-4690-3-1.

CHAPTER 7

Ariful Islam and Des Tedford, Risk Determinants of Small and Medium-sized Manufacturing Enterprises (SMEs) - an Exploratory Study in New Zealand, doi: 10.1186/2251-712X-8-12.

Index